Young People, Ethics, and the New Digital Media

This report was made possible by grants from the John D. and Catherine T. MacArthur Foundation in connection with its grant making initiative on Digital Media and Learning. For more information on the initiative visit www.macfound.org.

Young People, Ethics, and the New Digital Media: A Synthesis from the GoodPlay Project

Carrie James with Katie Davis, Andrea Flores, John M. Francis,
Lindsay Pettingill, Margaret Rundle, and Howard Gardner

Project Zero
Harvard Graduate School of Education

The MIT Press
Cambridge, Massachusetts
London, England

© 2009 Massachusetts Institute of Technology

All rights reserved. No part of this book may be reproduced in any form by any electronic or mechanical means (including photocopying, recording, or information storage and retrieval) without permission in writing from the publisher.

For information about special quantity discounts, please email special_sales@mitpress.mit.edu.

This book was set in Stone Serif and Stone Sans by the MIT Press. Printed and bound in the United States of America.

Library of Congress Cataloging-in-Publication Data

James, Carrie.
Young people, ethics, and the new digital media : a synthesis from the goodplay project / Carrie James ; with Katie Davis . . . [et al.].
    p.   cm.—(The John D. and Catherine T. MacArthur Foundation reports on digital media and learning)
Includes bibliographical references.
ISBN 978-0-262-51363-0 (pbk. : alk. paper)
1. Information society—Moral and ethical aspects. 2. Digital media—Moral and ethical aspects. I. Title.
HM851.J36   2009   302.23'10835—dc22   2009014658

10   9   8   7   6   5   4   3   2   1

Contents

Series Foreword

The John D. and Catherine T. MacArthur Foundation Reports on Digital Media and Learning, published by the MIT Press, present findings from current research on how young people learn, play, socialize, and participate in civic life. The Reports result from research projects funded by the MacArthur Foundation as part of its $50 million initiative in digital media and learning. They are published openly online (as well as in print) in order to support broad dissemination and to stimulate further research in the field.

Acknowledgments

This project was made possible thanks to a grant from the John D. and Catherine T. MacArthur Foundation's digital media and learning initiative. We are thankful to Jessica Benjamin and Sam Gilbert for their helpful comments and excellent research support and to Linda Burch, Katie Clinton, Henry Jenkins, Barry Joseph, Elisabeth Soep, Margaret Weigel, and Connie Yowell for helpful comments on earlier drafts of this report. In preparation for writing this report, we consulted educators, academic experts, professionals in the new digital media industry, and youth participants. We are extremely grateful for the insights and stories that they shared with us. We also thank the following individuals:

Sonja Baumer, Digital Youth: Kids' Informal Learning with Digital Media Project, University of California, Berkeley, and University of Southern California, Los Angeles

Steve Bergen, chief information officer, The Chapin School, New York

Matteo Bittanti, Digital Youth: Kids' Informal Learning with Digital Media Project, University of California, Berkeley, and

University of Southern California, Los Angeles; PhD candidate, new technologies of communications program, Libera Università di Lingue e Comunicazioni, Milan, Italy

danah boyd, Digital Youth: Kids' Informal Learning with Digital Media Project, University of California, Berkeley, and University of Southern California, Los Angeles; Researcher, Microsoft Research; Fellow, Harvard Law School's Berkman Center for Internet and Society.

Don Buckley, director of information and communications technology, The School, Columbia University, New York

Lena Chen, "Sex and the Ivy" blog, http://sexandtheivy.com

Chris Dede, Timothy E. Wirth professor in learning technologies, Harvard Graduate School of Education, Cambridge, Massachusetts

Nate Dern, social networker and comedian, Brooklyn, New York

Judith Donath, associate professor of media arts and sciences, Asahi Broadcasting Corporation career development professor, and director of the Sociable Media Group at the MIT Media Lab, Cambridge, Massachusetts

Kevin Driscoll, Prospect Hill Academy, Cambridge, Massachusetts; cofounder, TeachForward

Megan Finn, Digital Youth: Kids' Informal Learning with Digital Media Project, University of California, Berkeley, and University of Southern California, Los Angeles; PhD candidate, School of Information, University of California, Berkeley

Tom Gerace, founder of and chief executive officer at Gather.com

Heather Horst, Digital Youth: Kids' Informal Learning with Digital Media Project, University of California, Berkeley, and University of Southern California, Los Angeles; postdoctoral researcher, Annenberg Center for Communication, University of Southern California, Los Angeles

Global Kids Digital Media Youth Advisory, New York

Barry Joseph, director of the Online Leadership Program, Global Kids, New York

Jason Kaufman, John L. Loeb associate professor of the social sciences and head tutor, Harvard University, Cambridge, Massachusetts

Peter Levine, director of the Center for Information and Research on Civic Learning and Engagement CIRCLE), Tufts University, Medford, Massachusetts

Kevin Lewis, PhD candidate, department of sociology, Harvard University, Cambridge, Massachusetts

Derek Lomas, graduate researcher and codirector at the Social Movement Laboratory, California Institute of Telecommunications and Information Technology, University of California, San Diego

Jed Miller, Internet director, Revenue Watch Institute, New York; previous director of Internet programs, American Civil Liberties Union, Washington, DC

Mitchel Resnick, LEGO Papert professor of learning research, MIT Media Lab, Cambridge, Massachusetts

Rafi Santo, program associate, Online Leadership Program, Global Kids, New York

Christo Sims, Digital Youth: Kids' Informal Learning with Digital Media Project, University of California, Berkeley, and University of Southern California, Los Angeles; PhD candidate, School of Information, University of California, Berkeley

Nick Summers, cofounder of the *IvyGate* blog, http://www.ivy gateblog.com/

Sarita Yardi, Digital Youth: Kids' Informal Learning with Digital Media Project, University of California, Berkeley, and University of Southern California, Los Angeles; PhD candidate, human-centered computing program, Georgia Institute of Technology, Atlanta

Jonathan Zittrain, professor of Internet governance and regulation, Oxford University; Jack N. and Lillian R. Berkman visiting professor for entrepreneurial legal studies, Harvard Law School, Cambridge, Massachusetts; and cofounder and faculty codirector, Harvard Berkman Center for Internet and Society, Cambridge, Massachusetts

Abstract

The new digital media are a frontier that is rich with opportunities and risks, particularly for young people. Through digital technologies, young people are participating in a range of activities, including social networking, blogging, vlogging, gaming, instant messaging, downloading music and other content, uploading and sharing their own creations, and collaborating with others in various ways.

In late 2006, our research team at Harvard Project Zero launched a three-year project funded by the MacArthur Foundation. The goals of the GoodPlay Project are twofold—(1) to investigate the ethical contours of the new digital media and (2) to create interventions to promote ethical thinking and, ideally, conduct. In the first year of the project, we conducted background research to determine the state of knowledge about digital ethics and youth and to prepare ourselves for our empirical study. This report describes our thinking in advance of beginning our empirical work. We expect to revisit the framework and arguments that are presented here after our empirical study is complete.

In this report, we explore the ethical fault lines that are raised by such digital pursuits. We argue that five key issues are at stake in the new media—identity, privacy, ownership and authorship, credibility, and participation. Drawing on evidence from informant interviews, emerging scholarship on new media, and theoretical insights from psychology, sociology, political science, and cultural studies, we explore the ways in which youth are redefining these five concepts as they engage with the new digital media. For each issue, we describe and compare offline and online understandings and then explore the particular ethical promises and perils that surface online.

We define *good play* as online conduct that is meaningful and engaging to the participant and is responsible to others in the community and society in which it is carried out. We argue that the new digital media, with all their participatory potentials, are a playground in which five factors contribute to the likelihood of good play—the technologies of the new digital media; related technical and new media literacies; person-centered factors, such as cognitive and moral development, beliefs, and values; peer cultures, both online and offline; and ethical supports, including the presence or absence of adult mentors and educational curricula. The proposed model sets the stage for an empirical study that will invite young people to share their personal stories of engagement with the new digital media.

Young People, Ethics, and the New Digital Media

Introduction

When *Time* magazine declared its 2006 person of the year to be "You" (Grossman 2006), the magazine was pointing to an undeniable reality: anyone with an Internet connection can be a reporter, political commentator, cultural critic, or media producer. Around the same time, media scholar H. Jenkins and colleagues (2006) published a white paper extolling the "participatory cultures" of creation and sharing, mentorship, and civic engagement that were emerging online, especially among teens. Although *Time* did not explicitly frame participation in the new media as a youth phenomenon, most of the fifteen "citizens of digital democracy" who were featured in its December 13 article (Grossman 2006) were under the age of thirty-five. And Jenkins et al. (2006) strongly suggest that young people are especially well-poised to take full advantage of Web 2.0. Indeed, many young people are using the digital media in impressive and socially responsible ways. Consider the following examples.

TVNewser

In 2004, Brian Stelter, then a sophomore communications major at Towson University, started a blog called "TVNewser" that provides an ongoing, detailed record of ratings, gossip, and events in the news media industry. Over the past three years, "TVNewser" has become a chief source of information for news industry executives. In fact, Stelter receives frequent calls from people like Jonathan Klein, president of CNN's national news division. His youth and lack of credentials notwithstanding, Stelter is considered an extremely credible source (Bosman 2006). After graduating from college, Stelter was hired as a media reporter for the *New York Times*.

Global Kids

Global Kids (http://www.globalkids.org) is a New York–based organization that is "committed to transforming urban youth into successful students as well as global and community leaders." In 2000, Global Kids launched an Online Leadership Program (OLP) through which youth simultaneously build technical, new media literacy, leadership, and civic engagement skills. Youth participants engage in online dialogues about civic issues, regularly post comments on a blog, learn to design educational games and digital films, and play an active role in Teen Second Life, including its youth summer camp, which brings them together online to educate one another about global issues, such as child sex trafficking.

Yet for every digital superkid and for every example of good citizenship online, there seem to be many more examples of (intentional or naïve) misuses—or at least ethically ambiguous uses—of digital media. Consider these examples.

Lonelygirl15

In June 2006, a series of video blogs posted on YouTube by a teenager called Lonelygirl15 began to capture a wide audience ("Lonelygirl15"

2007). The videos depicted a sixteen-year-old girl named Bree talking about her day-to-day existence, including her experiences being home-schooled and raised by strict, religious parents. After several months, Bree was revealed to be Jessica Rose, a twenty-something actress who was working with several filmmaker friends to produce the video series (Heffernan and Zeller 2006).

The Digital Public

Aleksey Vayner, a senior at Yale University in 2006, became infamous after he submitted a résumé to the investment bank USB. Included with the résumé was his online, self-made video titled "Impossible Is Noth-ing," which appeared to be a record of Vayner's diverse talents and depicted him performing a variety of skills such as ballroom dancing and extreme weightlifting. The video link was circulated by email within the bank and soon beyond it. After it began making headlines in the blogosphere and in major newspapers, questions were raised about the authenticity of some of the footage. Vayner subsequently sought legal advice for what he considered to be an invasion of privacy (de la Merced 2006).

Speech in the Blogosphere

On April 6, 2007, a technical writer and prominent blogger, Kathy Si-erra, published an entry on her blog entitled "Death Threats against Bloggers Are NOT 'Protected Speech.'" For several weeks, Sierra had received anonymous violent comments and death threats on her own blog and on two other blogs. Following Sierra's alarming post, a heated controversy about the ethics of speech unfolded in the blogosphere. Calls for a blogger's code of conduct were met with angry protests that indicated how deeply many participants cherish the openness and free-doms of cyberspace (Pilkington 2007).

Ever since digital technologies were made widely available, scholars, educators, policymakers, and parents have been debating their implications for young people's literacy, attention spans, social tolerance, and propensity for aggression. Considerable strides are now being made in scholarship in many of these areas. The educational benefits of video games, for example, are being convincingly documented by scholars such as Gee (2003), Johnson (2005), and Shaffer (2006). At the same time, debates persist about the relationship between video games and violence (Anderson et al. 2004; Gentile, Lynch, Linder, and Walsh 2004).

Concerns about ethical issues in the new media have also been expressed by journalists, politicians, ideologues, and educators but have received less attention from scholars. In response to concerns about online predators, illegal downloading, and imprudent posting of content online, a number of cybersafety initiatives have emerged online and in schools around the country. The Ad Council's YouTube videos entitled "Think before You Post" seek to "to make teen girls aware of the potential dangers of sharing and posting personal information online and of communicating with unfamiliar people to help reduce their risk of sexual victimization and abduction" (Ad Council 2007). Youth-driven outreach groups and anticyberbullying campaigns, such as Teenangels and StandUp!, are making their way into schools. Somewhat surprisingly though, objective, research-based accounts of the ethical issues raised by the new digital media are scarce.[1] This report attempts to fill this gap.

Some of the digital media's ethical fault lines that we have scrutinized are the nature of personal identities that are being formed online; the fate of personal privacy in an environment where diverse types of information can be gleaned and disseminated; the meaning of authorship in spaces where multiple, anonymous contributors produce knowledge; the status of intellectual and other forms of property that are easily accessible by a broad public; the ways in which individuals (both known and anonymous) interact and treat one another in cyberspace; and the credibility and trustworthiness of individuals, organizations, and causes that are regularly trafficking on the Internet. We believe that five core issues are salient in the new media—identity, privacy, ownership and authorship, credibility, and participation. These issues have long been considered important offline as well. Yet in digital spaces, these issues may carry new or at least distinct ethical stakes. It thus seems critical to ask whether the new digital media are giving rise to new mental models—new "ethical minds"—with respect to identity, privacy, ownership and authorship, credibility, and participation and whether the new digital media require a reconceptualization of these issues and the ethical potentials they carry. As a starting point for considering these questions, we explore emerging data regarding how young people manage these five issues as they participate in virtual spaces. Our account considers the unique affordances inherent in the new digital media, and associated promises and perils are illustrated through each section's vignettes. The five themes explored here are ethically significant in the digital age, but they are not necessarily the final defining ethical fault lines of this age. We expect that our subsequent empirical work will turn up new ethical issues and perhaps suggest different ways of understanding these themes and the relationships among them.

A note about terminology: in this report, we use the term *new digital media* (NDM) or simply *new media* to refer to the actual technologies that people use to connect with one another—including mobile phones, personal digital assistants (PDAs), game consoles, and computers connected to the Internet. Through these technologies, young people are participating in a range of activities, including social networking, blogging, vlogging, gaming, instant messaging, downloading music and other content, uploading and sharing their creations, and collaborating with others in various ways (see appendix A for a detailed overview of youth involvement in specific digital activities). Of principal interest to us are those activities that are interactive (such as multiplayer as opposed single-player games), dialogical (online deliberation on Gather.com, for example), and participatory (user-contributed content, such as videos posted on YouTube). We use the terms *cyberspace, the Internet,* or simply *online* to denote the virtual realm in which such interactive activities are taking place. We also use the term *Web 2.0,* which refers to the second-generation Internet technologies that permit, indeed invite, people to create, share, and modify online content (O'Reilly 2005).

New Digital Frontiers

The new digital media have ushered in a new and essentially unlimited set of frontiers (Gardner 2007b). Frontiers are open spaces: they often lack comprehensive and well-enforced rules and regulations and thus harbor both tremendous promises and significant perils. On the promising side, the new digital media permit and encourage "participatory cultures." As Henry Jenkins and colleagues define it, "a participatory culture is a culture with relatively low barriers to artistic expression and civic engagement, strong support for creating and sharing one's creations, and some type of informal mentorship whereby what is known by the most experienced is passed along to novices. A participatory culture is also one in which members believe their contributions matter, and feel some degree of social connection with one another (at the least they care what other people think about what they have created)" (Jenkins et al. 2006, 3).

Time's 2006 person of the year points to the power of Jenkins's concept and suggests that the potential of the new media to empower ordinary citizens and consumers is being realized. Many cultural critics and social scientists (Jenkins among them) have argued that audiences of traditional media have never been passive (Lembo 2000; Radway 1985). Yet the new media invite a different level of agency. Blogs allow people to speak out about issues they care about, massive multiplayer online games invite players to modify them as they play, and social networking sites permit participants to forge new connections with people beyond their real-world cliques, schools, communities, and even countries. In the most idealistic terms, the new digital media hold great potential for facilitating civil society, civic engagement, and democratic participation (Ito 2004; Jenkins 2006a; Jenkins et al. 2006; Moore 2003; Pettingill 2007). If leveraged properly, the Internet can be a powerful tool for promoting social responsibility. At the same time, technologies themselves may be used for a range of purposes. The new media's capacities to promote evil might be in equal proportion to their capacities to promote good (Williams 1974). Indeed, the frontierlike quality of the new digital media means that opportunities for ethical lapses abound. There are innumerable ways— some barely conceivable—for the dishonest to perpetrate harms and, in turn, for the innocent to be victimized.

The potentials and perils of the new digital media are reflected in opposing discourses described as "digital faith" and "moral panics" (Green and Hannon 2007). Optimist Moore (2003) points to the "worldwide peace campaign" of millions of interconnected people who are working for social issues and human rights as a "beautiful" example of "emergent democracy" in cyberspace, while skeptic Keen describes the Internet as "a chaotic human arrangement with few, if any, formal social pacts. Today's Internet resembles a state of nature—Hobbes' dystopia rather than Rousseau's idyll" (2007, 2). These disputes echo those that have raged for decades (if not longer) about traditional media, especially with respect to effects on children (Buckingham 2000). Yet the new media may pose qualitatively different risks and opportunities. The reality is that most online situations are rich with promises and risks, both of which often carry ethical consequences.

Like all frontiers, cyberspace will eventually be regulated in some fashion, but it is unclear how regulation will occur and who will gain and who will lose from the regulation. The Blogger's Code of Conduct (2007) and the Deleting Online Predators Act (2006) are recent efforts in the direction of regulation that take two different tacks. The former, created by bloggers themselves, establishes guidelines for conduct; the latter, a bill introduced by legislators, restricts young people's access to social networking and other interactive sites. Moreover, because commercial interests have an ever-growing presence in digital spaces, the extent to which market forces will have a hand in regulation and the ethical implications of their involvement need to be considered. Now is the time to ask what a regulated World Wide Web would look like and how we can retain the openness and socially positive potentials of the new digital media while restraining unethical conduct. We believe that such a balance cannot be struck without a nuanced understanding of the distinct ethical fault lines in these rapidly evolving frontiers. Yet understanding is but a first step. Ultimately, for the promises of the new digital media to be positively realized, supports for ethical participation—indeed for the creation of "ethical minds" (Gardner 2007a)—must emerge.

In late 2006, our research team at Harvard Project Zero launched a three-year project funded by the MacArthur Foundation. The goals of the GoodPlay Project are twofold—(1) to investigate the ethical contours of the new digital media and (2) to create interventions to promote ethical thinking and conduct. In the first year of the project, we conducted background research to determine the state of knowledge about digital ethics and youth and to prepare ourselves for our empirical study. This report describes our thinking in advance of beginning our empirical work. We expect to revisit the framework and arguments presented here after our empirical study is complete.

Again, our objective in this report is to provide an overview of what is known about ethical issues that are raised by the new digital media, especially with respect to young people. We are motivated in our project by our concerns about the prevalence of ideologically driven (as opposed to empirically based) accounts of youth's online activities. Therefore, we strive to provide a balanced account that counters both disempowering skepticism of the new media and its opposite—uncritical celebration or "digital faith" (Green and Hannon 2007). In writing this report, we have three further goals—(1) to stimulate conversations with informed readers, scholars, and other critical thinkers about digital media; (2) to establish a research agenda to help confirm, reject, or revise the understandings and hypotheses presented here; (3) to provide hints about the kinds of supports needed (that is, the key ingredients for successful outreach efforts) so that young people can reflect on the ethical implications of their online activities and ultimately engage in "good play."

Note

1.   Exceptions include UNESCO's 2007 report, *Ethical Implications of Emerging Technologies* (Rundle and Conley 2007). The report presents the potential positive and negative effects of technologies such as the semantic Web, digital identity management, biometrics, radio frequency identification, grid computing, and other technologies that are now being developed or adopted. By contrast, this report explores the broad issues that are suggested by the activities occurring through media technologies that are widely available and frequently used, particularly by young people. See also the Vatican's 2002 report on ethics and the Internet title "Ethics in Internet": http://www.vatican.va/roman_curia/ pontifical_councils/pccs/documents/rc_pc_pccs_doc_20020228_ethics -internet_en.html.

The "Good Play" Approach

In this report, our understanding of what constitutes an ethical issue is deliberately broad and includes respect and disrespect, morality and immorality, individual behavior, role fulfillment, and positive (civic engagement) and negative (deception and plagiarism) behaviors. In setting out to explore young people's activities in the new media, voluntary leisure-time activities or play are foremost in our analysis, although work activities (such as schoolwork, research, and job seeking) are also carried out online by youth. As in the physical world, play in the new media includes gaming, but we also include activities such as instant messaging, social networking on Facebook and MySpace, participation in fan fiction groups, blogging, and content creation (including video sharing through sites such as YouTube). Many of these leisure-time activities fall arguably in a grey area between work and play. For example, blogging can be instrumental and goal-directed, constitute training for jobs, and lead directly to paid work. Our conception of play encompasses such activities because they often start out as hobbies that are undertaken in informal, "third spaces" without the support and constraints of (adult) supervisors, without rewards from teachers, and without explicit standards of conduct and quality. Much of our attention in this report is focused on these third-space activities and less so on unambiguous games. In labeling such activities *play*, we do not suggest that they are inconsequential. Rather, we do so to highlight the nature of the contexts in which they are carried out and the varied purposes that participants can bring to them.

We come to this effort after spending ten years researching good work—work that is excellent in quality, meaningful to its practitioners, and ethical (Gardner, Csikszentmihalyi, and Damon 2001). Among many relevant findings from this research is the discovery that good work and bad work are much easier to define and determine in professions that have explicit missions, goals, and values around which key stakeholders align. For example, it is relatively easy to detect when a physician is adhering to medicine's codes of conduct and mission because these codes are explicit, as are the outcomes of violations (such as high rates of patient mortality). It is more difficult to delineate good work in business or in the arts because these are relatively unregulated spheres of work. Journalism lies somewhere in between a bona fide profession and an unlicensed, unregulated sphere of work.

The ethics of play may be even more difficult to discern because (depending on the activity) participants do not necessarily come to it with consensual goals and values. Play can be experienced by players as both "utterly absorbing" and yet low stakes—"a free activity standing quite consciously outside 'ordinary' life" and, by implication, "outside morals" (Huizinga 1955,

13). At the same time, play needs to be taken seriously because it expresses important cultural mores. As Geertz (1972) so convincingly argued, play (particularly "deep play") emerges from and serves as a "metasocial commentary" on the culture in which it occurs. At the same time, some players have much greater appreciation of the make-believe and metacognitive aspects of play (Bateson 1972). All aspects of play do not harbor ethical implications, but many do, and greater awareness of their ethical potentials is surely warranted.

Play in the new digital media is fraught with different (and perhaps greater) ethical potentials and perils than offline play because participants can be anonymous, assume a fictional identity, and exit voluntary communities, games, and cyberworlds whenever they please. In short, accountability depends on the strength of ties within a given online community; where ties are weak, accountability may be rare. At the same time, online play is carried out in a digital public before a sometimes vast and unknowable audience so that a young person's YouTube mash-up can begin as a fun after-school activity and in short order become the object of ridicule or even a spark for serious political deliberation around the world. Because so much online activity is proactive or constructionist—creating content, sharing content, or simply crafting online identities through profiles (Floridi and Sanders 2005)—a significant onus is placed on creators to consider the broad implications of their actions. Moreover, although conscious perpetrators and clear victims of misconduct surely exist at play, unintentional lapses may be more commonplace. For example, Aleksey Vayner, described in this report's opening vignette, surely never imagined that his video résumé would be scrutinized and mocked by a vast public. Because well-intentioned acts may result in significant, unintended harms, clear perpetrators and victims may not easily be discerned. Understanding the ethics of play is thus more urgent and yet may be more difficult than studying the ethical facets of good work. To guide our efforts, we rely on the following conceptual anchors:

• **Respect and ethics**   Our principal focus is ethics, but this discussion also considers its close ally, respect. The distinction between the two concepts is worth noting. As we define it, *respect* involves openness to differences, tolerance of others, and civility toward people, whether or not they are personally known. The respectful person gives others the benefit of the doubt. Respect or disrespect can be observed by and directed toward very young children and will soon be recognized as such. In contrast, *ethics* presupposes the capacity for thinking in abstract terms about the implications of a given course of action for one's self, group, profession, community, nation, and world. For example, "I am a reporter. What are my rights and responsibilities?" or "I am a citizen of Boston. What are my rights and responsibilities?" Ethical conduct is closely aligned with the responsibilities to and for others that are attached to one's role in a given context.

• **Roles and responsibilities**   At the heart of ethics is responsibility to others with whom one interacts through various roles, including student, athlete, worker, professional, community resident, citizen, parent, and friend. Such roles can be transposed to new media activities where youth are game players (akin to the athlete or team member role), online community members (citizens), bloggers (writers or citizen journalists), and social networkers (friends). (See appendix A for a detailed overview of the range of roles that young people are assuming online.) Regardless of the context (offline or online, social or work), ethics are part of one's membership in a group, the roles that one assumes, and the responsibilities that are stated or implied therein.

▪ **Emic and etic** The distinction between emic (internal) and etic (external) is taken from anthropology and linguistics. It allows us to distinguish between an individual's phenomenological experience and a trained observer's interpretations of her words and actions. Young people may not have an emic (internal) awareness of themselves as playing out various roles, offline and online. However, from an etic (external) perspective, they are assuming roles as students, employees at work, and children to their parents; such roles carry implicit, if not explicit, responsibilities. Accordingly, online conduct can have broad consequences that are not easily grasped by young people and are not transparent to them as they blog, post photos and videos on MySpace and YouTube, and interact with known or unknown others in virtual worlds such as Second Life.

▪ **Good play** Accordingly, we define *good play* as online conduct that is both meaningful and engaging to the participant and responsible to others in the community in which it is carried out. We consider how and why identity, privacy, ownership and authorship, credibility, and participation are managed in responsible or irresponsible ways by youth in online contexts. Again, definitions of responsible or ethical conduct in online spaces may differ markedly from offline definitions. Here we consider the new digital media as a playground in which the following factors contribute to the likelihood of good play—(1) technical literacy and technology availability; (2) cognitive and moral person-centered factors (including developmental capacities, beliefs, and values); (3) online and offline peer cultures; and (4) presence or absence of ethical supports (including adult or peer mentors, educational curricula, and explicit or implicit codes of conduct in digital spaces). Our approach to ethics does not focus solely on transgressions but strives to understand why, how, and where good play happens. We therefore delineate both perils and promises in the new media. Like new media literacy advocates (Buckingham 2003; Jenkins 2006a, 2006b; Jenkins et al. 2006; Livingstone 2002), we wish to move beyond naive optimism or pessimism and encourage critical reflection on the considerable variation in the purposes and values that young people bring to their online activities.

In the analysis that follows, we explore the ethical implications—both positive and negative—of the various activities in the new media in which young people in particular are engaged. We draw on evidence from over thirty interviews with informants, including academic experts, industry representatives, educators incorporating the new media into their curricula, and youth who are especially engaged in some aspect of the new media. Interviews were approximately one hour in length, semi-structured, and partially tailored to each informant's specific area of expertise. Questions focused on the broad opportunities and challenges of the new media, youth trends in online participation (both positive and negative), and specific ethical dilemmas that have come up in each informant's teaching, research, new media work, or online participation (see appendix B for standard interview protocols). We also draw on the growing literature on games, social networking sites, blogs, knowledge communities, and civic engagement in cyberspace, as well as long-standing research and theory about youth, media, and culture.

Several limitations in the nature of evidence that we draw on are worth noting. First, our data rely heavily on adult informants and scholarship. Second, the handful of youth informants with whom we spoke are highly engaged with the new media, often assuming leadership roles in online communities, games, and blogs. For these reasons, their perspectives may not be representative of the average young person.

Digital Youth

The headlines with which we began this report touch on the ethical issues that surface online but also refer to typical online pursuits of "digital natives" (Prensky 2001)—people who have grown up around and who regularly engage with new media. As the Berkman Center's Digital Natives Project aptly points out, not all youth are "digital natives," nor are all "digital natives" young people (Digital Natives 2007). Yet our attention here focuses on that intersection of youth and digital fluency. We believe that the promises and perils of the new media are especially salient for those young people who possess digital skills, spend considerable amounts of time online, and are assuming new kinds of roles there. These young people may be best prepared to use new media for good but may also be the most likely perpetrators or victims of ethical lapses. Our interviews with informants suggest that young people are often confused by the power of new technologies and easily do things (like download music and copy and paste images, text, and software) that are technically illegal and may be ethically questionable. Because of their technical skills, a leader of a digital youth group calls young people today "babies with superpowers": they can do many things but don't necessarily understand what their actions mean and what effects those actions can have.

Indeed, psychological research on moral development suggests that capacities for moral decision making and action evolve over time and are affected by social contexts and experiences (Kohlberg 1981; Turiel 2006). At the same time, most research on moral development focuses on individual decisions with reference to other persons in their world. There is much less known about the evolution of moral or ethical stances in public spheres like interactive media or in relationships with institutions. As youth participate in digital publics at ever-younger ages, questions about their developmental capacities (what we might expect of young people at ages fourteen, eighteen, and twenty-five?) seem particularly important when considering their capacities for discerning the ethical stakes at play in the new digital media. Traditional psychological frameworks of moral development may need to be revised in light of the distinct properties of digital media and young people's heavy participation with them from early ages.

To start, we need to consider evidence regarding how young people conceive of the ethical responsibilities that accompany their new media play. Do young people hold distinct conceptions of their responsibilities and of the key ethical issues at stake in their online pursuits? Many informants with whom we spoke claimed that digital youth are qualitatively different from older generations in an ethical sense. Awareness of ethical implications of online conduct is reported to be generally low, although variation is acknowledged. As one researcher put it, youth can range from "completely delusional" to "hyperaware" of the potential audiences. More generally, the young are purported to have distinct ethical stances on core issues such as identity, privacy, ownership and authorship, credibility, and participation. One educator also noted that young people frequently assume that all participants share the same ethical codes, even though ethics are rarely explicit online.

In the account that follows, we draw on these impressions of the ethical stances of digital youth by asking how and why traditional stances on such issues might be challenged in digital contexts. At the same time, we treat them as hypotheses to be explored through further empirical research.

Ethical Fault Lines in the New Digital Media

Are youth redefining identity, privacy, ownership, credibility, and participation as they engage with the new digital media? If so, how, why, and with what consequences? Drawing on insights from interviews and relevant literatures, we address these five issues below. For each issue, we begin with a fictionalized vignette that highlights the key ethical fault lines that we believe are at play. We then compare traditional (offline) conceptions of each issue with evidence of new (online) conceptions of the issue and explore the distinct promises and perils of online conceptions.

The order in which we address these five issues is deliberate: we begin with the self and then move outward to the self's relationships with objects, with other persons, and with society. We explore identity (the ego itself and how one's self is represented and managed online), privacy (one's choices about disclosure of personal information in the digital public), ownership and authorship (one's relation to objects, including intellectual property), credibility (one's trustworthiness and assessment of others online), and participation (one's social relations, conduct, and membership in broader communities).[1]

1. Identity

Identity Play on MySpace

Zoe is a sixteen-year-old high-school honors student who is shy but has a small circle of good friends. Like many teens, Zoe has a MySpace page. When she first joined MySpace, her parents expressed concern about stories that they had read about adult predators and reckless online conduct by youth. After some debate, Zoe persuaded them to allow her to remain on MySpace but had to grant them access to her page. After a few months, this arrangement began to feel stifling, and so, without telling her parents, Zoe created a second MySpace identity named Zee, age eighteen.

Zoe uses her Zee page to write more openly about her feelings and experiences and to explore alternative identities. In designing her Zee profile, Zoe posts pictures of "herself" that are actually photos of a long-time friend from summer camp whom she considers to be more attractive and older looking than she is. After all, she figures, her Zee profile is more of a play space, and the odds that her friend will find out are slim, especially because they are in touch only rarely. Zee makes a number of new, online friends, including Dominick, whose profile states that he is twenty years old and lives in a nearby town. Zoe begins an online relationship with Dominick as Zee, who behaves more flirtatiously than Zoe. She finds her interactions with Dominick thrilling and enjoys the opportunity to perform as a more assertive identity. After several weeks of flirtation, Dominick proposes that they meet offline. Zoe is flattered but wonders how he will react when he meets her and learns that the photos on her page are not of herself.

Questions raised: How can online self-expression and exploration play a positive role in a young person's identity formation?

Under what circumstances does identity play become deception? What do young people gain when they deliberately and strategically perform their identities in a public space? What are the potential costs to both themselves and others?

Identity Play, Offline and Online

Theorists of human development have described identity formation as the major task of adolescence, at least in modern Western societies (Erikson 1968). During this period in their lives, individuals begin to reconsider their conceptions of themselves as they become increasingly aware of the broader society, including its values, norms, and expectations. Psychologists have identified exploration as the key mechanism through which adolescents can try on different identities and experience how they are received by society (Moshman 2005; Schwartz 2001). Erikson (1980) thus described adolescence as a "psychosocial moratorium," a "time out" that allows youth to experiment freely with their identities in a low-stakes environment. Ideally, this experimentation results in an identity that makes sense to both the individual *and* to society. As Erikson notes, identity formation "is dependent on the process by which a *society* (often through subsocieties) *identifies the young individual*," a process that begins in adolescence but recurs throughout an individual's lifetime (1980, 122). The social nature of identity is further underscored by symbolic interactionists who argue that the self develops and is continually enacted and reshaped in a social context (Cooley 1902; Goffman 1959; Mead 1934). With respect to our purposes here, identity formation is not just an individual project but a deeply social one that hinges on social validation, carries social consequences, and bears ethical promises and risks.

Identity exploration and formation are facilitated by self-expression, self-reflection, and feedback from others. Offline, young people explore their identities in a variety of ways. They may experiment with clothing and hairstyles, adopt the attitudes of music or other subcultures, or become involved in extracurricular activities that develop a talent, a passion, or an ideology. They can engage in self-reflection through solitary journaling and can elicit feedback in face-to-face interactions with friends, known peers, and adults. However, offline identity explorations are constrained in a number of ways. For instance, individuals cannot easily change the shape or size of their bodies. Youth are also limited by the opportunities and social roles that are made available to them. A boy will have difficulty trying on the role of dancer if there are no dance classes in his neighborhood or if his family and friends believe that men should not be dancers. Similarly, a girl may feel that she cannot reveal her assertive side if the adults in her life value female submissiveness. As these examples suggest, feedback from others is a critically important source of validation (or, in these cases, repudiation) of one's identity experiments. Offline, feedback is typically received from close relations, including friends, peers, and family, which can be limiting.

Not only are young people limited by the types of identities that they can explore offline, but the spaces and times that are available to them for exploration may be disappearing. Adolescence today involves more pressures—related to schoolwork, extracurricular activities, and college admissions—than it did when Erikson first described the adolescent moratorium or when Hall (1904) first wrote about adolescence a century ago. According to Turkle (1999), the moratorium is being cut short by the high-stakes pressures that today's youth face. Adolescents have decreasing amounts of time and space in which explore their identities.

At the same time, the new media are providing adolescents with new spaces for identity exploration. Indeed, Turkle (1999) has described the Internet as a fertile space for youth to undertake Erikson's psychosocial moratorium. Freed from the physical, social, and economic constraints of real life, she argues, individuals can experiment with multiple identities in an environment that is perceived to be "low-stakes." Turkle's pioneering book (1995) described how individuals engage in identity play on the Internet by adopting different names, writing styles, and personas for their digital "selves." More than a decade later, the number and types of digital spaces have expanded, making it possible for many more forms of self-expression and spaces for self-reflection to emerge. Young people can thus elicit feedback on their identity experiments from broader, more diverse audiences than they can offline. Although opportunities to adopt radically different identities exist in many online spaces, researchers are finding that youth's online self-expressions tend to reflect aspects of their offline selves (Huffaker 2006; Valentine and Holloway 2002). Youth use their MySpace pages, Facebook profiles, and blogs to express their values and cultural tastes, sexual identities, personalities, and feelings about their relationships and experiences. These online expressions are necessarily more deliberate than offline ones. As boyd (2007b) points out, online youth have to write themselves into being.

The real developmental task of identity formation is increasingly happening in virtual spaces. It is therefore critical to consider the implications of these new social contexts for the kinds of identities that are explored and formed and their effects on others. Again, identity formation is undertaken by individuals, but it both affects and is affected by relationships with others, pushing it squarely into ethical terrain. Also, actions to the self can be considered ethical or unethical if the self is understood as a role that one assumes. In this section (and in the privacy section below), we consider identity play's ethical promises and perils with respect to the self and mainly interpersonal relationships. In the sections that follow, identity play resurfaces as we consider broader opportunities and risks online, such as those related to ownership and authorship, credibility, and participation in communities.

The Promises of Virtual Identity Play

Virtual identity play can aid the identity-formation process by providing new tools and diverse spaces for self-expression, self-reflection, and feedback from others. First, online spaces offer multiple avenues for creative self-expression or identity play. Zoe can customize her MySpace page by choosing certain colors, design motifs, and music; by posting pictures, poetry, and song lyrics; and by making lists of her favorite bands, movies, and books. On her Zee page and blog, Zoe expresses her feelings and the aspects of her personality (such as assertiveness, candor, and sexuality) that her shyness prevents her from conveying to others in the physical world. Indeed, Zee's expressions and the ways that she interacts with others may be authentic representations of Zoe's self or of a "possible self" that Zoe is consciously forming and aspires to achieve in the real world (Markus and Nurius 1986). Because the stakes may be perceived to be low, online spaces (especially anonymous or semianonymous ones) may be treated as "safe" places to explore identities, work through personal issues, or even "act out" unresolved conflicts with others (Bradley 2005; Turkle 2004). Zoe could even extend her identity experimentation further by constructing an avatar in Second Life and exploring sexual flirtations with women in a more anonymous way. Such opportunities to "take the role of the other" (Mead 1934) can help Zoe figure out both who she is and wants to be and may engender greater appreciation for the perspectives of others, possibly increasing social tolerance and respect. This ability to place oneself in another's shoes is one prerequisite for ethical thinking and conduct.

Second, the need to write one's online identity into existence (boyd 2007b) can encourage self-reflection, and reflection can nurture greater awareness of one's roles and responsibilities to oneself, to others, and to one's community. To reconcile one's childhood roles with the roles that are made available and valued by society, an individual must engage in a certain amount of self-reflection. Stern (2007) suggests that the deliberate nature of online self-representations facilitates identity formation by forcing individuals to articulate who they are now, who they want to become, and what beliefs and values guide them in their personal growth. At the most concrete level, Zoe defines her online self through the pictures that she posts, the lists of favorites (bands, movies, and books) that she creates, and the personal information (such as name, age, and geographic location) that she chooses to share. On a more abstract level, Zoe has the opportunity, through her blog entries, to reflect on how her experiences and interactions relate to each other and have meaning for her and others. Through her identity experiments, Zoe may be pushed to consider consciously what kinds of responsibilities are implied by enacting a given identity and whether an identity is aligned with her responsibilities to her self (including her beliefs and values and the person that she aspires to become). Moreover, Zoe may also consider whether her expressions align with the expectations of others (including her parents and online friends such as Dominick) and her responsibilities to them. Self-reflection is an important personal skill that facilitates broader social and ethical skills and can help engender credibility and socially responsible participation (as we discuss in separate sections below).

Finally, online spaces provide youth with unique and important opportunities to gain validating feedback from others. Human development occurs in a social context and is aided by feedback that helps individuals reconcile their self-conceptions with society's appraisals of them. Stern (2007) describes the value to adolescents of the feedback that they receive online. She observes that online spaces offer adolescents an opportunity to have a voice, an opportunity that may be rarer offline. Moreover, youth can test and receive feedback on different versions of themselves, such as their sexuality or unexplored aspects of their personality. If the feedback that they receive is positive, then they may feel more confident about integrating these versions into their offline identities. If the feedback is negative, then they have the chance to revise their identities as many times as they wish without embarrassment or disgrace offline— provided that their online experiments are undetected by others. In Zoe's case, she gained confidence from the positive feedback that she received when she adopted a more assertive and flirtatious identity on her MySpace page. Moreover, as their own selves are validated, youth may be better poised to extend themselves to and validate others. Social validation, which is increasingly attained online, may prevent social alienation and disaffection and social harms such as bullying, hate speech, and violence.

The Perils of Virtual Identity Play

Although identity play through the new digital media can be beneficial, the forms of self-expression, self-reflection, and feedback that are conducted online may undermine, rather than enhance, an individual's identity formation. Young people who fail to develop a coherent, autonomous sense of self are evading an important obligation to themselves. They may struggle in myriad ways and be incapable of assuming important social roles and fulfilling responsibilities. After all, as noted before, identity formation is a social process: its successes and failures affect others, sometimes in negative ways. Social harms can result when identity experimentation crosses over to deception and when explicitly harmful identities are explored. Additional perils to the self and more indirect harms to others arise when youth's identities become deeply fragmented, when self-reflection is overshadowed by self-promotion, or when youth become overly dependent on feedback from others. Our focus here is mainly on the self and known others, yet identity lapses can have broader consequences online, at times harming numerous unknown and distant others (Silverstone 2007).

In relationships with others, identity play can easily cross over to deception. Online friends and strangers can be misled about the nature of a person's offline identity. Even in contexts such as Second Life, where identity experimentation is expected and promoted, avatars can develop online relationships and mislead others about characteristics of their offline selves (such as their sex, age, or sexuality). The extent to which such information is deceptive or merely part of the play depends on the expectations of the individuals involved, and those expectations are rarely explicit. When Zoe (as Zee) began an online flirtation with Dominick, she didn't expect that the relationship would have an offline dimension, so she didn't feel compelled to say that Zee's photos were part of the play and not of her true self. In certain cases, offline friends who know about a young person's online life can become concerned and confused by a growing disjuncture between online and offline personas. For example, morose poetry or song lyrics that are posted on a young person's MySpace page can signal underlying struggles on the part of a young person and an earnest attempt to reach out to others. At worst, a LiveJournal blog can be a deceptive performance that is aimed at garnering attention, as in the fake deaths that have been concocted by troubled youth online (Swains 2007).

Related to this, spaces such as Second Life and massively multiplayer online games (MMOGs) may permit youth to explore harmful identities, such as rapist, murderer, or misogynist, although for the perpetrator, the potential real-world effects of engaging in online rape and hate speech are contested (Anderson et al. 2004; Gentile, Lynch, Linder, and Walsh 2004; Lynn 2007). For those who are the focus of aggressive or even violent writings, virtual acts can create offline feelings of intimidation and fear (as evidenced by blogger Kathy Sierra's experience), yet little consensus exists among adults about what is appropriate decorum (Pilkington 2007). Many online communities, such as fan cultures, have entrenched codes of ethics that are supported by strong ties between participants (Jenkins 2006b). However, newcomers to these communities and other online spaces may find that the proper limits of identity play are less clear, making young people vulnerable to aggression and unintentional lapses.

Evidence collected to date suggests that most young people's online identities reflect key elements of their offline identities (Huffaker 2006; Valentine and Holloway 2002). However, concerns have been raised about youth who experiment with radically different identities (Glass 1993; Turkle 1995). Turkle notes that "without any principle of coherence, the self spins in all directions. Multiplicity is not viable if it means shifting among personalities that cannot communicate" (1995, 58). Placing this risk in the context of Zoe's vignette, her Zee MySpace identity could become increasingly unrelated to her offline self. If at school Zoe remains a somewhat shy but easygoing and friendly person and online she expresses the more assertive and sexual aspects of herself, she may feel increasingly frustrated if she can't connect her online and offline selves. According to Erikson (1968), the ultimate goal of an adolescent's identity explorations is a coherent, unitary sense of self, not a series of fragmented identities. At this point, the long-term effects of online identity play are unclear, and researchers need to explore how it facilitates positive exploration and poses obstacles to establishing a healthy sense of self (Buckingham 2007). Ultimately, though, Erikson's conception of a healthy self may need to be reconsidered in light of the new opportunities for identity development that are provided by the new digital media.

Another possible peril of online identity play lies in its performative quality. The self-reflection that digital spaces afford can be undermined when presenting to an audience becomes more valued and urgent than turning inward to engage in self-examination. Goffman (1959) uses the metaphor of a theater to describe the ways in which people relate to one another as actors in a staged play, and the performative element of self-presentation may be magnified online. For example, Stern (2007) interviewed teens who constructed their personal homepages and blogs in a deliberate and strategic way. Using cultural artifacts, they crafted their online identities with an eye toward attracting and entertaining a public audience. They omitted the parts of themselves that did not fit their desired performance and augmented the parts that did. On her Zee page, Zoe selected certain photos (in this case, not her own), colors, and music to present a specific identity to her online audience. Her performance may be personally meaningful, but it is nevertheless directed outward and shaped by external cultural symbols. It seems reasonable to question the degree to which one can engage in deep and genuine self-reflection while spending a great deal of energy performing a specific self to others. The performance also risks becoming more important than the truth. A blogger who chronicles his sexual exploits can exaggerate for the sake of creating a compelling story and in so doing might depict friends, peers, colleagues, and others in a negative light, which raises privacy issues, which we address below.

Finally, although online spaces allow adolescents' ideas, self-expressions, and confessions to be shared with others, the feedback that they seek and receive can be problematic. Opportunities for disclosure can set the stage for an overreliance on feedback, which can undercut autonomy and create fragile identities. The recent case of Megan Meier, a thirteen-year-old girl who committed suicide after an online "friend" began to taunt her, is an extreme example of this peril. Turkle (2008) uses the term "tethering" to describe the nearly constant connectivity to others and sharing of information that is permitted, indeed encouraged, in the digital media. Mobile technologies and "status" modules on instant message programs and social networking sites are commonly used by young people to signal their current locations, activities, and even moods to their online networks. Feedback is encouraged and even expected from others. When young people are encouraged to maintain continuous connections with others and to express and reflect in a fully or semi-public space, the benefits of autonomous self-reflection—indeed, of "being alone"—come to be undervalued. Young people may be developing an unhealthy reliance on feedback from others as a basis for self-development and limiting their capacity for autonomous decision making (Moser 2007; Zaslow 2007). In turn, a strong desire for positive feedback and praise from others might interfere with a young person's capacity for reflecting in an abstract, disinterested way about the ethical implications of his or her conduct.

The Ethics of Online Identities

Virtual identity play may provide youth with unique opportunities to develop healthy identities, but this outcome is by no means guaranteed. Under the best of circumstances, young people are able to express different aspects of themselves in a supportive environment, engage in self-reflection, and elicit constructive feedback from others. However, the new media can also pose significant risks to a young person's sense of self, including risks related to identity deception, opportunities to assume (or be attacked by) harmful virtual identities, and an unhealthy reliance on feedback and connectivity to others. Further research is needed to uncover the conditions under which digital participation facilitates and detracts from the development of healthy, autonomous, and socially responsible identities.

2. Privacy

Privacy in the Blogosphere

Sofia is an eighteen-year-old freshman at a small college. She has been keeping a blog on LiveJournal for several years and has continued to blog after she enters college to keep her high school friends informed of the ups and downs of her new life at college. She also finds that writing is a great way to think through problems in her life and to express her opinions in a free environment. The stresses of taking premed courses, handling dorm life, and making new friends are consistent themes of

Sofia's blog at college, but she also writes about her dating and intimate experiences. Some aspects of her posts are fictionalized, but Sofia has fun writing, and judging by the comments that her friends leave on her blog, they seem to enjoy her narratives. She has told only a couple of close friends at college about her blog and disguises the identities of her crushes, hook-ups, and dates. Although she does refer to her college by name, she writes under a pseudonym and doesn't give many details that would clearly identify her as the author. Even if a few random people happen across her blog, she reflects, they probably wouldn't be able to figure out her real identity.

A local journalist who is writing a story on blogging searches LiveJournal for local college students who actively maintain blogs. Her search uncovers Sofia's blog, and because it is in the public domain, the journalist feels free to write about its content. After the story appears in the local newspaper, Sofia is surprised to find that students all over campus start reading and commenting on her blog. Eventually, a few people are able to piece together details from her posts and expose Sofia as the author of the blog. Some of her past romantic partners express anger and frustration because comments that Sofia's friends write on the blog reveal their identities. Sofia feels blind-sided by this turn of events. She never imagined that a broader public would be reading about her most intimate thoughts and experiences.

Questions raised: What does it mean to manage online privacy in an ethical manner? How do online spaces facilitate and undermine ethical thinking about privacy? How much personal information is reasonable to share online? Are young people who share personal experiences online taking steps to protect their own and others' identities, and are these steps sufficient? Is it reasonable for young people to expect a certain measure of privacy when it comes to their online lives? Who is at fault when an unintended audience can read a young person's revealing blog or MySpace page? What might be the long-term offline consequences for the blogger?

Privacy, Offline and Online

Privacy refers to how a person's personal data and information about others are handled in social contexts. Offline, *privacy* is generally understood to mean the retention or concealment of personal information, and in the United States, it is framed as an entitlement. The private is kept to oneself or shared only with close, trusted, face-to-face friends. The *right to privacy* is frequently invoked to protect sensitive information (such as an individual's finances, medical history, and intimate relations) from public view. Modern privacy statutes in the United States, which were first advocated by Warren and Brandeis (1890) in the late nineteenth century, reflect a desire to protect individuals from exposure to the public through the press and from unwarranted search and surveillance by the state. The right to be "let alone" and the right of the individual to maintain freedom from authoritative institutions are the main concerns of current legal statutes dealing with privacy offline (Woo 2006).

Because the new digital media allow personal information to be shared with a broad public, they are making privacy issues more salient and at the same time altering conventional understandings of privacy. Nonintervention by institutions is still a concern and is perhaps heightened by the new media. Yet distinct properties of the Internet bear on privacy in new ways. boyd (2007a) identifies four such properties, including persistence (what you post persists indefinitely), searchability (you can search for anyone and find their digital "body"), replicability (you can copy and paste information from one context to another), and invisible audiences (you can never be sure who your audience is). Despite these features, many young people share deeply personal information with one another on sites such as MySpace, LiveJournal, and Facebook, and much of this information is (or easily can become) accessible by a broad public.

The sharing that is happening in these spaces does not necessarily suggest that youth do not value their own privacy or respect others' privacy, but it does suggest that they understand privacy differently than earlier generations did. To many young participants, privacy is not about hiding personal information but rather involves carefully managing its disclosure—what is shared, how it is presented, and who can access it (Woo 2006). Online, young people are arguably creating a *culture of disclosure*, meaning distinct beliefs, norms, and practices that are related to their online profiles and lives. This culture legitimates and guides young people's disclosure of personal information for their intended audiences of peers. For example, on sites such as LiveJournal, MySpace, and Facebook, a young person carefully chooses which personal details to disclose and how public to make this information. Choices may be based on the norms of the space (gleaned through studying the disclosing patterns of their peers), personal goals (to meet new friends, communicate with offline friends, or form a fan group), and beliefs (naïve or realistic) about a potential audience.

Online, a number of strategies—including privacy settings, selective disclosure, code switching, and deception—are used by youth to control the presentation of their identities and thereby manage their privacy. Most social networking sites have privacy settings that allow users to limit access to their profiles to a narrow audience of confirmed friends, and evidence suggests that many young people use them. According to a recent Pew Internet and American Life Project survey, 66 percent of participants in teen social networking sites report restricting access to their profiles in some way (Lenhart and Madden 2007). Participants can also use selective disclosure: they fill out only a portion of the fields provided by the site to indicate their personal information, often omitting details like last name, city of residence, and so on. In another strategy, users may have one social networking profile to interact with friends and another less detailed or partly fictitious profile to interact with strangers. An educator with whom we spoke called this practice *code switching* and noted that it provides a sense of control by allowing people to present different identities in different contexts. Finally, deception is a widely used practice for enhancing online privacy. According to Pew, among teens whose profiles are public, 46 percent say they give at least some false information (Lenhart and Madden 2007). Taken together, these privacy strategies can produce either multiple identities or one fragmented identity, both of which can preserve a sense of privacy while still allowing for disclosure and participation.

The prevalence of privacy strategies suggests that online privacy is being consciously managed by many young people. But this is only part of the story. Other evidence suggests that some youth (and adults) fail or choose not to use protection strategies. Pew reports that Internet users are becoming more aware of their "digital footprints" but that surprisingly few of them use strategies to limit access to their information (Madden et al. 2007). This laissez-faire approach to disclosure can be interpreted in a number of ways, ranging from carelessness to a conscious (although fragile) set of assumptions and norms about an audience. In imagining their audiences, many young people (perhaps naively or egocentrically) assume that only invited friends will read their profiles or blogs and that the uninvited (such as parents and teachers) will respect their privacy and treat their online expressions as if they were off limits, as a handwritten journal would be (boyd 2007b; Weber 2006). As Weber (2006) says, "public is the new private: young people often realize that their blogs and homepages are public and accessible, but they trust that only their peers are interested enough to view them. Adults are supposed to know where they are not welcome and act accordingly." Normative codes among youth participants (for example, the belief that information that is shared online should not be copied and pasted into another context without permission) may also contribute to lack of use of privacy strategies. In our vignette, Sofia assumed that the small circle of friends who knew about her blog would not refer to it or paste content from it onto their Facebook pages. The conception of privacy here shifts the responsibility for ethical management of personal information away from the writer to the audience, the scale of which, whether acknowledged or not, is often inherently unknowable.

The Promises of Online Privacy

The online culture of disclosure holds important promises for young people, including empowerment of themselves and others, the creation of communities of support around shared struggles, and the development of a broad ethical sense of responsibility with respect to privacy.

As noted in our treatment of identity above, online communities are fertile spaces for identity development because they encourage self-expression, self-reflection, and feedback. Most relevant to privacy is how online disclosure can be carried out (partially or fully) anonymously and yield positive, comforting, and even empowering feedback. Young people can feel empowered by the ability to tell their stories and reflect on struggles in their lives online through blogs, Facebook, and virtual worlds. Sofia found her voice as a writer through her blog and gained insights into herself and her relationships with others through writing about them. Positive feedback from her readers increased her confidence and encouraged her to continue to write. Sofia's blogging could open future doors for her in journalism or fiction writing. Furthermore, Sofia's reflections could inspire other young women (especially women in restrictive family or school situations) to express themselves and their sexuality. Posting their stories and reflections in the digital public, young women may unintentionally be doing a kind of consciousness raising that is similar to what second-wave feminists did through print books in the 1960s.

Online disclosure of personal stories can also yield support for troubled youth. Social networking sites such as MySpace and Facebook invite young people to reveal private aspects of themselves with strangers and build communities around common struggles. Young people who are struggling with sensitive issues can reach out anonymously to find support for personal problems that they may fear disclosing face to face, forming support groups around issues such as how to reveal gay or lesbian sexuality to families, cope with shyness, and stop practicing self-injury (cutting). Practices that allow youth to present fragmented identities (such as anonymous participation, code switching, and deception) can help youth build communities of support while maintaining a sense of control over sensitive information. A teen's anonymous online journal about his struggles to grow up in an alcoholic family could become an important source of comfort, support, and perhaps even action for other young people in comparable family situations. Young people also use online communities to cope with tragedies, as was demonstrated by the many online memorials that were written after the 2007 shootings at Virginia Polytechnic Institute.

A final promise of the online culture of disclosure is that some young people develop a genuine ethics of privacy that helps them present themselves and handle other people's information in a considerate and responsible way. Many youth who disclose personal information online assume that their audience will behave responsibly. Such assumptions can be naïve and expose youth to significant risks, but if made explicit, they could help youth and online communities be guided by an ethics of responsibility.

The Perils of Online Privacy

The potential perils of the culture of disclosure are numerous. Youth can harm themselves and others by failing to understand the persistence of, searchability of, replicability of, and invisible audiences for the information that they share about themselves online. Deception that is intended to protect oneself can also have unintended negative effects.

The fragile assumptions that are made by young people like Sofia about other bloggers and about audiences for their online identities can create significant risks. Sofia's intended audience for her personal reflections was her close friends. She assumed that others who came across her blog would click the Back button out of respect for her privacy. This assumption was shattered when a journalist wrote an article that placed Sofia and the people about whom she wrote in an uncomfortable and potentially damaging position. Even though she took some measures to control her online identity, Sofia was caught off guard and thrust into the public eye. Sofia's reputation as a friend, classmate, and responsible writer was damaged by her failure to consider fully the risks and responsibilities of anonymous or selective disclosure in a digital public. Furthermore, Sophia's blog entries may have harmed her subjects—the romantic partners and friends about whom she wrote—in unpredictable ways ranging from their reputations at school to their future opportunities beyond it.

Indeed, unwitting participants in the digital public may be the most frequent victims of privacy lapses. According to boyd (2007b), many young people develop MySpace and Facebook profiles for other people so that they can add those names to their own profiles and thus exaggerate the extent of their offline popularity. Some young people are therefore yielding control over the creation of their online identities to their friends, with little understanding of the broader, potentially negative consequences. A semifictionalized blog entry about a friend's predilection for shoplifting or a video from a party posted on YouTube can negatively affect another person's reputation and opportunities for the indefinite future if accessed by unintended viewers such as college admissions officers and potential employers. This risk of "collapsed context" (boyd and Heer 2006) is particularly worrisome in the social networking environment. Even a nonincriminating photograph or video of a young athlete stretching before a track meet can have unforeseen negative effects if it is posted online. Recently, a fan posted the photograph of an attractive high school student on a football message board, the photo was discovered by a sports blogger with a wide audience, and the image was spread across the Internet. Within days, a YouTube video showing the student at a track meet was viewed over 150,000 times (Saslow 2007), and the student was experiencing online harassment. This story suggests that a person's identity, reputation, and sense of safety in the world may be increasingly beyond her control as the new media permit rapid and widespread sharing of information. This story also highlights the responsibilities that young people have to one another to handle the personal information and content they disclose to each other online with care. Overall, the culture of disclosure works if all potential audiences operate with the same ethical code regarding access and use of the information that is available online.

The final privacy-related peril relates to deception. If online deception is done to protect the writer's privacy, it is largely viewed as proper and is even encouraged by many parents. According to this view, deception can be a safe way to participate online. The lack of face-to-face interaction makes deception easier online than in real life. Nevertheless, the line between benign and malicious deception can be difficult for young people to discern in mediated spaces where outcomes are not immediately clear. For example, pretending to be someone you are not online (such as an expert in a field, a potential friend, or a potential romantic partner) can harm others, even if the harms are distant or invisible to the perpetrator (Silverstone 2007). Furthermore, as boyd (2007a) has suggested, it is worth considering the broader message that is being conveyed to young people when they are encouraged to misrepresent themselves online, even if it is for safety's sake. Decades ago, Bok (1979) argued that profound societal harms—such as the decline of pervasive trust—are associated with habits of lying. The great potentials of the Internet will not be realized if basic trust cannot be forged among participants.

Related to this is the unknowable distance between a young person's online identity and an audience. This is a perilous feature of the new digital media (Silverstone 2007). Privacy strategies such as code switching and deception perpetuate the problem of unknowable social and geographic distance between online participants. What results, according to Silverstone (2007, 172), is a "polarization. . . . The unfamiliar is either pushed to a point beyond strangeness, beyond humanity; or it is drawn so close as to become indistinguishable from ourselves." Both scenarios pose risks and set the stage for ethical misconduct. For example, the culture of online disclosure can cause young people to form potentially dangerous relationships with other users, as in cases of teen-adult predator relationships. At the same time, anonymity and deception can reduce accountability in online spaces and lead to online aggression, ranging from griefing (intentionally irritating participants in an online game or other community) to anonymous death threats (as in the attacks on blogger Kathy Sierra).

The Ethics of Online Privacy

An emergent culture of disclosure in the new digital media holds both risks and opportunities for young people. On the one hand, carefully managed and informed sharing can inspire and empower youth, build supportive communities for the troubled, and encourage an ethics of privacy in others. On the other hand, careless oversharing can have long-term negative effects on young people and the friends about whom they write and whose online identities they cocreate. Deception for safety's sake can also create confusion and pose risks. The promises of the digital public can be realized and the perils avoided, however, if young people consider the implications of their self-presentations in light of the properties of persistence, searchability, replicability, and invisible audiences (boyd 2007b) that characterize the new media. For Silverstone (2007, 172), "proper distance" emerges from the "search for enough knowledge and understanding of the other person or the other culture to enable responsibility and care. . . . We need to be close, but not too close, distant but not too distant." An integral part of this "proper distance" is modulating the sharing of personal information—preserving a sense of individual privacy while maintaining openness to community. Future studies are needed to confirm or revise hypotheses about digital youth's mental models of privacy. We need to understand the extent to which their approaches are distinct from offline models, consciously formed, and considerate of the promises and risks of engagement in the digital public.

3. Ownership and Authorship

Authorship in Knowledge Communities

Daniel is a high-school senior who is interested in social movements and occasionally contributes articles to Wikipedia, the online encyclopedia. When he is asked to write a research paper about an American protest movement for an American history class, he decides to write about the immigration rallies that took place in several American cities on May 1, 2006. In his paper, he draws extensively from a Wikipedia entry about the rallies to which he contributed a few months earlier. After reading Daniel's paper, his teacher calls him into her office and accuses him of plagiarism, noting that he used verbatim lines from Wikipedia without giving proper credit to the source. Daniel replies that since he was a contributor to the Wikipedia article, his use does not constitute plagiarism. He also argues that the passages he used were mainly historical supporting facts and that the core of the paper is his unique analysis of the rallies' significance as a protest movement. Above all, he asserts, the purpose of Wikipedia is to make knowledge available for widespread use. It does not provide the names of article authors, and he will not be cited by others for his contributions. In fact, authorship is irrelevant.

Questions raised: What perils are associated with the free flow of information online? What does *authorship* mean in knowledge communities like Wikipedia, and what constitutes fair use of articles on the site? To whom should writers give credit when citing information from knowledge communities, and who is the victim when credit is not given? How might exposure to and participation in online knowledge communities be engendering a new ethics of authorship? On the whole, how are concepts of ownership changing in the new digital media?

Ownership and Authorship, Offline and Online

Offline authorship and ownership are tied to the legal concept of property (intellectual or tangible), which gives the ownership and exclusive intellectual property rights for a work to an individual or organization. In short, credit and profit are given to creators or owners. In schools, plagiarism codes help guide students about the fair use of offline copyrighted materials and citation styles. Most universities have strict antiplagiarism and peer-to-peer (P2P) laws but also retain constitutional and contractual rights to intellectual freedom and freedom of information (Putter 2006). Offline ownership and authorship are well-defined concepts that are protected by law and reinforced by cultural norms in corporations and schools.

The offline stability of these concepts does not mean that violations of fair authorship and ownership do not occur there. Our previous data (Fischman, Solomon, Greenspan, and Gardner 2004) suggest that pressures to succeed, poor peer norms, and an absence of mentors contribute to offline transgressions. According to a recent report from the Josephson Institute of Ethics (2006), 60 percent of high-school-age youth admit to having cheated on a school test, almost 30 percent to having stolen from a store, and 33 percent to having plagiarized from the Internet for an assignment, providing further evidence that participating in a "cheating culture" (Callahan 2004) may be routine for many youth. So even though offline authorship and ownership are well-protected, clearly defined legal concepts, lapses are still fairly commonplace.

For various reasons, online ownership and authorship are less clear than their offline versions. Technology allows users to copy and paste copyrighted materials. With this widespread availability of pay-for-use versions and free Internet content, software, and files, determining what is freely available and what is not can be confusing. This confusion may be accompanied by younger users' naïve belief that if something is downloadable, then everybody can use it without payment. Lenhart and Madden (2005) report that "teens who get music files online believe it's unrealistic to expect people to self-regulate and avoid free downloading and file-sharing altogether." At the same time, online applications, such as wikis and hosted documents, are making authorship and ownership increasingly collaborative and are blurring the distinctions between author and audience. Comments on a blog affect the content of a blogger's entries, and a gamer's changes to a game's code are used by the company in next-generation versions. Finally, in contrast to offline legal restrictions, attempts to regulate online intellectual property rights and copyright through Digital Rights Management and the Digital Millennium Copyright Act of 1998 have proven difficult to enforce.

These features of ownership and authorship in the new media are influencing ethical stances on these issues, and distinct cultural norms and attitudes are developing regarding online materials, particularly among young people. Daniel's justification for his failure to cite Wikipedia in his paper suggests a nontraditional understanding of authorship. An educator with whom we spoke reported that, thanks to the digital media, young people live in and embrace an "infringing culture" where they expect immediate access to information and goods. Considered in a more positive light, a new ethics of "free culture" (Lessig 2004) and collaboration may be emergent. Either way, implications for creators and lawful owners of music, video, images, and text are uncertain. What is clear is that past conceptions of ownership, authorship, and copyright are now contested and are likely to be significantly revised or reinterpreted for the digital age.

Promises of Ownership and Authorship Online

Much critical attention has focused on online transgressions, but the new conceptions of ownership and authorship that are emerging online offer significant promise for young people. Increasing opportunities for cocreation (Jenkins 2006a) and participation in "knowledge communities" (Lévy 1999) can provide youth with new skills that can empower them to become engaged citizens and successful workers.

The new digital media shift the traditional separation between—and roles and responsibilities of—audience and author, forging opportunities for cocreation that may be especially advantageous to youth. Cocreation of content can include writing fan fiction in an online community and contributing new code to preexisting commercial games. Opportunities for cocreation grew exponentially with the advent of Web 2.0. For example, one prominent youth blogger with whom we spoke noted how his readers became cocreating tipsters regarding facts and stories. The virtual world Second Life has an open code model that allows users to build their own modifications to the world while retaining authorship and ownership benefits. Although cocreation can produce tensions regarding an author's obligation to an audience or a company's rights in its game, virtual world, or site, these practices also allow readers and players empowering opportunities to assume creator and contributor roles.

Participation in cocreation can build new skills, efficacy, and empowerment. On a more abstract level, Web 2.0 demystifies authorship and ownership for youth and invites them to see themselves as creators and active participants in something larger than themselves. Gamers who create new levels in games or modify their avatars may be prompted to consider future careers that they may not have thought possible, such as software engineering. Cocreation allows users to create their own dynamic works, moving beyond passive modes of entertainment to active engagement with texts. Would-be journalists can practice their narrative and editorial skills through blogging and posting comments on others' blogs, while aspiring filmmakers can post their serial minidramas on YouTube. Such experiences can be considered practice for adopting future professional roles as producers and can carry stakes that are commensurate to those that accompany professional work. The stakes that are associated with cocreation in a digital public, in participatory cultures, or in other online "affinity spaces" (Gee 2004) can push a young person to consider her role as a creator and the responsibilities that this role implies.

A second promise of the new media with respect to authorship and ownership is open access to knowledge and information. Open-source advocates argue the virtues of providing content and free information to the masses and invite their contributions to production and design (Lessig 2004). The open-source movement promotes the idea that sharing information may lead to higher-quality creations, greater knowledge, and more efficient knowledge-building processes. In their roles as students and learners, wired young people are poised to be the main beneficiaries of this exciting democratization of knowledge. Young people Google facts that they hear on television, rely on Really Simple Syndication (RSS) readers or aggregators for the latest news, and, like Daniel, find background information on Wikipedia for school assignments. With freedom of information, youth have access to vast resources for learning, experience rich intellectual exchanges, and connect to knowledge as never before possible. Implications for their future roles as workers and citizens are stunning.

Freedom of information, if handled properly, can engender a deep respect for the work of others. Use of Creative Commons licenses provides an excellent model for online intellectual property law that both protects the creator and keeps quality work accessible to the public (http://creativecommons.org). If youth are taught to use these new authorship and information paradigms, perhaps they will share their works with others and take part in knowledge communities. Moreover, as information is freely available, a democratizing effect could create new opportunities for civic engagement for the individual and community. Daniel's role as a Wikipedia contributor could spark his interest in participating in protest movements offline and inspire him to take an active citizen role. Freedom of information and increased interactivity with texts destabilize the traditional conceptions of authorship, ownership, and the roles of authors and audience, but this destabilization lowers the barrier of participation for youth, generating more active and critically engaged young users who are empowered to act rather than just watch or react.

Like Daniel, young people can feel empowered when they contribute their expertise to knowledge communities. Some educators are experimenting with class assignments that ask students to contribute to Wikipedia (BBC News 2007). On a concrete level, they can gain valuable skills such as teamwork. On a more abstract level, they can learn to appreciate the importance of respect and ethics in collaboration. Moreover, personal responsibility can be cultivated through online knowledge communities where youth are expected to contribute meaningfully. Knowledge communities may actually serve as an antidote to plagiarism, some informants suggested, by simply providing "many eyes on the work" and by increasing students' awareness of their responsibilities to one another. Sharing work in progress online (through class wikis, for example) can help build students' skills in peer critiquing, knowledge building, and grasping the meaning of quality work.

Some benefits can be derived from antiplagiarism communities such as Turnitin.com, but the most important promise of knowledge communities is not in identifying bad play but rather in advancing good play and learning. Cocreation and knowledge communities provide youth the opportunities to assume the roles of creators and collaborators, learn the responsibilities that are associated with these roles, and build valuable skills for their futures as workers and citizens.

The Perils of Ownership and Authorship Online

The perils that can arise around ownership and authorship are numerous. Youth can be at risk of exploitation by corporate entities, can abuse information and content (as in illegal file

sharing and downloading), and can be confused about authorship distinctions in knowledge communities.

Young people's authorship and ownership claims can (and
often do) go unacknowledged when they cocreate online. For
example, gamers and game companies have a symbiotic relationship, and yet intellectual property rights typically lie with
the companies. As noted by Postigo (2003), hobbyist game modders (modifiers) cast a spotlight on the sometimes contested
nature of ownership and authorship in the games space. Modders are gamers who hack into game code and create new game
play levels, new elements of the virtual world, and other game
play components for no monetary reward. Although modders
are infringing on the copyrighted materials of game companies,
the companies benefit greatly from modders' innovations. This
activity produces extended game play, fan bases, and design
ideas for new products while decreasing development time and
labor costs. Yet most modders are denied authorship credentials,
compensation, and ownership rights and are sometimes pejora-
tively labeled *copyright infringers* or *hackers*. This exploitation is
not limited to game modders. The online creations of aspiring
filmmakers, musicians, and writers can just as easily be misap-
propriated by corporate interests and other users or through
restrictive licensing agreements. As stated in the "MySpace
Terms of Service" (MySpace 2007), users who publish content
thereby grant to MySpace a "limited license to use, modify, publicly perform, publicly display, reproduce, and distribute such
Content solely on and through the MySpace Services." The
advertisements that are posted beside youth content on My-
Space, YouTube, and other sites yield profits for the site owners

that are not shared with youth creators. Exploitation of young people's play-work may lead youth to embrace a "free culture" approach (Lessig 2004), but it is also possible that youth will come to hold little regard for the integrity of their own and other's work and to deny the responsibilities to others that are implied in cocreation.

A second ethical peril for youth is the temptation to abuse the free flow of information and content online. With music, video, and other content, the law dictates that copyright-protected materials cannot be widely distributed without purchase. Corporations representing musicians, for example, are working feverishly to manage rampant illegal downloading, as in the current legal battles between the Recording Industry Association of America and college students over illegal music downloads. The prevalence of illegal downloading suggests that young people's conception of ownership might be that they are entitled to what they can easily access online. As one educator we interviewed put it, young people perceive "no sense of scarcity" in the virtual world.

Setting aside the question of legality, it is important to consider where and when such appropriation is clearly unethical and where it is arguably appropriate, even ethical. Green and Jenkins (2009) use the concept of "moral economy" from Thompson (1971) to capture the ways in which music downloaders and fans justify their appropriation and repurposing of content. The average young person (or older person, for that matter) may be unlikely to perceive how illegal downloading victimizes mammoth entertainment companies or celebrity entertainers. Indeed, one informant noted that youth often frame their illegal downloading or file sharing in Robin Hood–like terms, referring to the concentration of wealth and power by large media companies and producers: "Artist X is already wealthy, so my illegal download doesn't matter." Moreover, a recent survey of European youth reported that low levels of trust in entertainment companies may be an important factor that contributes to piracy (Edelman 2007). A different stance is often held by participants in fan communities. Green and Jenkins (2009) suggest that fans' remixes of copyrighted content can actually increase the visibility, popularity, and success of the original content, yielding great benefits in the long run for media producers and owners. In short, fans create more fans, who then purchase original content.

The conflicting stances of different stakeholder communities suggest that the ethics of music downloading and appropriation are far from clear. In keeping with our conception of ethical conduct, consumers who are capable of thinking in abstract terms about their responsibilities to others (and not simply about their own interests) are engaging in ethical thinking. Are music downloaders and fans fully considering the perspectives of media producers and owners? In turn, are media producers and owners considering the stances of users and remixers? What constitutes ethical appropriation in this situation? Definitive answers to these questions are beyond the scope of this report, but a failure to forge consensus over these issues may be problematic for all stakeholders. Further research is needed to shed fuller light on the beliefs and ideas held by youth—and adults—with respect to these issues.

The downloading and appropriation of young consumers and cocreators might represent an ethical (even if not legal) stance, but one peril for users is that a sense of entitlement becomes a habit of mind that is overextended to other contexts. In school work, appropriation without giving credit to original authors can constitute clear academic dishonesty. The extent to which ethical mental models regarding some forms of appropriation cross over to other forms is unknown but appears to be an important question for further research. Daniel's mental model regarding uses of Wikipedia is probably affected by his role as a contributor there, but it may also stem from (and cross over to) his experiences with other forms of media. As a contributor to Wikipedia, he holds certain beliefs about the knowledge that is built and shared on the site and expectations about its appropriate uses. Nevertheless, Daniel's standpoint might be at odds with that of other contributors, who may see Wikipedia as the work product of dedicated individuals who deserve credit. If Daniel is also a hobbyist game modder in his spare time and feels exploited by commercial game owners, he might come to see the Internet as a free-for-all. In addition, Daniel might observe adults around him engaging in a range of ethically questionable practices, such as software piracy, without an explicit or coherent justification. Although online plagiarism, illegal downloading, and software piracy are widely discussed as youth transgressions, adult participants can add to the confusion. On the whole, it seems clear that young people may be deprived of opportunities for learning about the perspectives of different stakeholders and reflecting on the ethics of appropriation of online content.

Finally, in the realm of authorship, what happens to credit in an era of knowledge communities and collaborative work? Some individual creators may want (and need) credit for their work to satisfy personal pride, to demonstrate competence and achievement, and to make a living. Others, like Daniel, may consider it irrelevant. Conceptions of authorship and responsibilities to authors may be unclear to many youth users and participants in knowledge communities. As a Wikipedia contributor, Daniel felt responsibility to the knowledge reproduced there and was not concerned with giving or collecting credit. Yet traditional educational institutions still operate on the single-author model, with implications for citation norms and notions of fair use. Teachers who maintain traditional notions of authorship and credit and who punish students for treating material from Wikipedia differently may miss opportunities to engage their students about evolving notions of authorship. As Davidson (2007) notes, recent criticisms of Wikipedia, such as the 2007 decision of the history department at Middlebury College to ban its use, overlook the great opportunity that such sites provide for teaching research methods and credibility-assessment skills.

The Ethics of Ownership and Authorship Online

In an age of file sharing and knowledge communities, ownership and authorship have become muddy issues. Young people and all other new media users are caught between old and new modes of authorship and ownership. Confusion about what constitutes ethical appropriation and what contesting notions of authorship are held by different stakeholders may be on the rise. The worst-case scenario is that youth will embrace an overreaching sense of entitlement with respect to knowledge and other creations in digital circulation. In their future roles as workers, they may avoid teamwork for fear of not receiving due credit or perhaps be apt to usurp their colleagues' products as their own. Conversely, the same youth could become tomorrow's innovators, pooling their skills, talents, and resources for the greater good. Crucial to these promising outcomes is fostering productive dialogue among teachers and students about authorship, ownership, and fair use in a digital age. As conceptions of authorship, ownership, and the responsibilities that are implied in each are destabilized, building consensus around new conceptions of these issues and revising old conceptions for the digital age are priorities for both youth and adults.

4. Credibility

Expertise and Credibility in Online Forums

Maya is a twenty-four-year-old receptionist at the local gym, where all employees receive basic training in cardiopulmonary resuscitation (CPR) and emergency treatment for injuries. Maya observes the trainers in the gym closely and notes the kinds of workouts that they suggest for their clients. She has been interested in fitness and health since an early age and keeps up on the latest exercise and diet information by reading magazines and visiting GetTrim.com, a social networking site about healthy living where experts and nonexperts interact.

Maya notices that some participants on GetTrim.com report difficulties improving their health, and she feels sure that she can help them. She posts that she is a state-certified trainer and an expert in health and fitness and is available to share her knowledge with the community. A few users seek out Maya's advice on various exercise and nutrition matters and begin her suggested regimes. Within a few weeks, users are posting their positive results and encourage others to contact Maya. Soon Maya is giving advice to many GetTrim.com users on a wide range of issues.

Josh, who is one of the master trainers at the gym, decides to advertise his services as a personal trainer on GetTrim.com. He notices that many users are talking about Maya's advice, so he checks out her profile. To his surprise, he discovers that she is the gym receptionist and claims to be a state-certified expert. Josh confronts Maya in the online community forum about her lack of credentials. Maya does not respond to Josh's comments. Josh then makes a complaint to the site administrators, who close Maya's account due to a breach of the site's rule about truthful representation. This triggers a heated exchange among Maya's satisfied clients, members of the community who are genuinely certified, and those who are outraged by her deception.

Questions raised: What role do offline credentials play in online credibility? Can deception about credentials harm the cohesion of online communities? Why might someone misrepresent his or her expertise online? What harm can be done and to whom?

Credibility, Offline and Online

We consider two faces of credibility here—the ways in which young people establish their own credibility and young people's capacities for assessing the credibility of others. Although the ability to evaluate others' credibility is important and can have ethical implications, our principal concerns here are the judgments and actions of young people that affect their own credibility. How do young people decide to present themselves—their credentials, skills, and motivations—to various others in various contexts? For our purposes here, credibility is about being accurate and authentic when representing one's competence and motivations.

Offline credibility is typically conveyed through credentials, which are achieved through education, certification, on-the-job training, and a reputation for competence. Credentials take time to accrue but, when achieved, reliably signal competence. But credibility is also determined by the integrity of a person's interests and motivations. A highly qualified and esteemed medical doctor who is touting a new drug may not be deemed credible if she is discovered to hold stock in the drug company. Her motivations can be called into question: is she promoting the drug because in her professional opinion it is effective or because she has a stake in the company's profits? Motivations can be difficult to discern, but they are an important aspect of credibility. In the vignette, Maya seems to have good intentions: she wants to share her knowledge to help others. However, she does not have the requisite qualifications to work as a trainer or to publish an article in a reputable health magazine. She has not yet established her credibility in the offline health and fitness world.

Credibility is relatively easy to define with respect to working adults. But what does credibility look like among young people who have not yet completed their education or entered the workforce? Youth signal their credibility in their everyday activities in various ways. In school, a young person demonstrates competence and good intentions by completing her school work diligently and competently and by achieving good grades without cutting corners. At home, she shows credibility by competently carrying out her chores and following (most of) her parents' rules. With her friends, she keeps trusted secrets, provides support, and follows through on her commitments. In her community, she volunteers at various events for the sake of the community. Across these contexts, credibility is achieved through a track record of fulfilling obligations competently and with clear and good intentions.

Certain qualities of the new media, particularly the absence of visual cues, affect how credibility is signaled and assessed online. The new media's hallmark "low barriers to participation" (Jenkins et al. 2006) mean that people with diverse backgrounds, competencies, and motivations—experts and nonexperts, honest persons and poseurs alike—can have a voice in a variety of online spaces. Depending on the context, verifying the credibility of participants can be important. When medical advice is dispensed, for example, presenting competence in a truthful way is critical. Credibility may be less (or at least differently) important in spaces explicitly designed for fantasy play, such as Second Life. Other key qualities of the new media that bear on how credibility is conveyed include the potential for anonymity, the asynchronous nature of communication online, the relative absence of mechanisms for accountability structures or authority figures and mentors, and the ephemeral nature of ties in many online communities.

Signaling credibility is at once easier and more difficult online when traditional means for conveying competence and motivations are unavailable. A young person can join innumerable online communities where credibility will be judged by the quality of his participation, including his conduct and creations.

He can contribute to Wikipedia, become a guild master in *World of Warcraft*, post an amateur music video on YouTube, join and lead a political discussion group on Gather.com, or start a blog about reproductive rights. Feedback from the community helps determine his credibility in these spaces. Maya joined an online community to share her knowledge and gained positive feedback and increasing requests for advice. In the online health and fitness world, she gained credibility through the quality of her contributions and their presumably positive impact on people's lives.

The Promises of Online Credibility

Online conceptions of credibility can hold distinct promises for young people and the online communities in which they participate. Youth can be empowered by opportunities to demonstrate expertise. Provided access to the Internet, anyone can participate in public online communities. Online communities can be "affinity spaces" (Gee 2004) where diverse participants collaborate around a shared purpose or interest with little concern for differences in age, gender, ethnicity, and other status markers. People are not barred from entry simply because they lack formal training and credentials. Young people can act as experts due to their competence alone. Dialoguing and cocreating on an equal playing field with adults, young people can experience "collegial pedagogy" (Chavez and Soep 2005). As noted above, Brian Stelter started his TVNewser blog as an undergraduate and now attracts a massive audience, including top news media executives. In short, fewer restrictions exist online about what counts as knowledge and who qualifies as an expert.

The openness of the new media permits young people to explore different domains and outlets for their skills without the costs and time that are usually associated with training and education. Blogging and game modding can be considered quasi-internships or apprenticeships that prepare youth to enter fields such as journalism and engineering, which they may have not considered before they began their online activities. Opportunities to interact and perhaps cocreate with individuals with greater knowledge and expertise may help engender subject-matter expertise, facilitate skill development, and nurture key interpersonal skills including teamwork. From an early age, the new media can offer youth opportunities to try out new roles that may prepare them to become adept professionals, collaborators, and citizens.

In turn, domains such as journalism, software engineering, game design, and civil society can benefit from the present and future contributions of many young cocreators. Online knowledge communities such as Wikipedia demonstrate these reciprocal benefits: young people feel empowered by the opportunity to contribute, and diverse contributors facilitate good knowledge building. Ideally, such experiences help engender in youth a broader perspective, a feeling of efficacy, and a sense of responsibility. The broad definitions of expertise and credibility that exist online can thus yield positive social outcomes for individuals, communities, and society as a whole.

The Perils of Online Credibility

Although the distinct ways in which credibility is granted online can be beneficial, they also provide numerous occasions for misdeeds, including opportunities for deception and misrepresentation of one's identity, competence, and motivations. The relative absence of online visual cues and visible accountability structures allow various forms of deception to flourish, making it difficult to ascertain the credibility of participants' claims. A person can readily post someone else's work as her own, pay for someone to advance her in a game, misrepresent herself as a professional, or join a voluntary community with the hidden intention to disrupt it or to promote disguised commercial interests. Certain qualities of new media thus make assessments of credibility qualitatively different and arguably more difficult than in offline situations.

Online, young people might feel tempted to misrepresent their identities (who they are, how old they are, where they are, or what they do) and their backgrounds (what they have done and what their skills and capabilities are) because identity verification is difficult. Online cues that signal one's credibility can be unreliable and misleading (Donath 1999). Maya falsely stated in her profile that she was state certified, and she didn't need to provide evidence to support her claim. Such misrepresentations also occur offline and can go unnoticed for decades, but accountability online may be even rarer.

The forms of identity experimentation that are encouraged in certain online spaces can contribute to an attitude that fictional identities are permitted in all kinds of online communities. This attitude can be problematic in spaces where one's offline identity, competence, and motivations genuinely matter—as on WebMD, where consumers expect to find articles about breast cancer treatments that are written by board-certified physicians and researchers. Credentials often serve valuable purposes in online spaces; they can reduce risks by providing security through a process of vetting. Children, and even some tweens, may not yet be equipped developmentally to differentiate between contexts in which identity play is acceptable and expected and those in which offline credentials need to be presented.

Maya's story highlights the potential disconnect and tensions between offline and online credibility. Offline, she was barred from helping gym members because she lacked credentials. Online, participation in GetTrim.com did not require explicit credentials: Maya could freely dispense advice and be judged by the quality of her contributions. Yet at the same time, offline understandings of credibility affected Maya's online conduct. Well aware of the certificate requirements of the gym, Maya believed that it was necessary to appear credentialed for GetTrim users to heed her advice, and online it was easy for her to misrepresent herself. Being transparent about the extent and limits of one's expertise therefore becomes critically important online.

Motives and goals are hard to ascertain online due to the ability to be anonymous, the superficiality of some online relationships, and transient membership in some online communities. Maya's motives seem to be harmless. She wants to help others by sharing her knowledge with the community; her intention was not to give false or dangerous information. However, others might have more sinister motives. A corporate representative could post an anonymous testimonial about a potentially dangerous weight-loss supplement on GetTrim.com, and visitors to the site would have no way to verify the validity of such claims.

With few accountability structures in place online, everyone is responsible for his or her self-representation. In this, the support and guidance of adult mentors could be beneficial to young people. However, the gulfs that exist between the average adult's understanding of the new media and the ways that young people engage with it may virtually preclude good mentoring. If the new digital media's savviest participants cannot find a way to manage credibility themselves, the broader peril is that external parties will regulate their participation—imposing restrictive rules, erecting barriers to access in many online spaces, and stifling participatory cultures.

A further peril that is associated with online credibility is that young people may begin to undervalue credentials and miss opportunities to gain valuable but less readily acquirable skills. If everyone can participate and pose as an expert, formal training and education may seem unnecessary. As more and more readers compliment her on the advice that she gives, Maya may begin to feel that she is as capable as the trainers in her gym and does not need to take classes and gain legitimate qualifications. Positive feedback from GetTrim.com users may lead her to overestimate her competencies and believe that credentials are irrelevant. Furthermore, as Maya's clients begin asking for advice on a broader range of issues outside of her knowledge base, she may feel compelled or entitled to respond. Overextending her areas of expertise, she risks giving harmful advice. She also risks doing irreparable harm to herself. After her deception is revealed to the digital public, it may haunt her for the indefinite future.

The Ethics of Online Credibility

Participatory cultures offer youth unparalleled opportunities to develop and demonstrate knowledge and skills, assume roles as leaders and experts, and thus earn credibility at a relatively early age. At the same time, the relative absence of accountability structures permits deception. The desire to participate in certain online spheres and the perception from the offline world that credentials matter might lead young people to misrepresent their qualifications. Even if well-intended, deception of this kind can pose risks to both deceiver and deceived. Genuine credibility is built on truthfulness and transparency about competence (and its limits) and motives. Young people who understand and fulfill the responsibilities that are implied when credibility is granted to them are more likely to retain and nurture it online and off.

5. Participation

Civic Engagement on YouTube

Xander is a twenty-two-year-old nature photographer who is interested in the environment and its sustainability. He belongs to a Google group that was started by other nature photographers. One day, a message is posted to the group about a YouTube competition on environmental stewardship for Earth Day. Xander checks out the site and notices that much of the material submitted to the competition is accusatory, places blame for environmental problems on politicians, and fails to note the everyday changes that people can make to help the environment. Xander decides to make a video montage of his nature photos and overlay it with statistics about climate change and suggestions on how to live green. He mentions this idea to some friends. One suggests that he alter some of his images with Photoshop because most viewers won't understand the wide extent of environmental damage unless the photos are dramatic. Although he agrees that more dramatic photos might affect the audience more deeply, Xander thinks that using Photoshop in this way is inappropriate. Instead, he gathers information from the Union of Concerned Scientists Web site and the Worldchanging blog for his video, which he cites in his submission.

After the launch of the video, members of the YouTube group comment on his artistic technique and the uplifting tone of his submission. An anonymous user leaves a comment accusing Xander of copying images from a popular nature Web site and falsifying statistics. Xander ignores the criticism, but the anonymous user returns and launches a defamatory attack on him. He is offended but confident about his work and chooses not to engage the offending commenter.

Questions raised: What is ethical participation in online communities? What standards of behavior on sites like YouTube guide youth conduct? What ethical codes guide how content is created and shared? Do the new media create new opportunities for civic engagement? In what ways are young people assuming the role of citizen online? Are distinct responsibilities implied by cybercitizenship?

Participation, Offline and Online

Participation is the culminating ethical issue in the new digital media, and it arguably subsumes the issues of identity, privacy, ownership and authorship, and credibility. Participation centers on the roles and responsibilities that an individual has in community, society, and the world. It takes various forms, including communication, creation, sharing, and use of knowledge and information in all spheres of life—political, economic, and social. For the purposes of this report, we consider three aspects of participation—(1) access to a given sphere and to the basic skills and roles that allow participation in it; (2) standards of behavior in a given sphere, including those related to speech and conceptions of fair play; and (3) proactive participation, such as content creation and civic engagement.

In offline social, economic, and political life, access to participation is often limited to those who have certain resources, credentials, and attributes (such as age, race, sex, class, geographical region, resources, and capital). Young people often have limited access to skills and to roles that permit a voice in key spheres of decision making (political, economic, educational), creation, and distribution of knowledge and information. Constraints limiting the diversity of participants in those spheres affect the kinds of issues that are raised, decisions that are made, and content or knowledge that is produced. In addition, implicit and explicit standards of behavior in formal spaces like schools are often created and enforced top down by those in power. In such settings, roles, responsibilities, and sanctions for rule violations are typically explicit. Finally, young people's amateur creations (such as writings, music, and photography) can be shared locally but are not easily distributed to a broader audience. In the offline civic realm, youth might not feel that they have a voice with regard to public issues.

Civic participation varies across historical periods and geographical regions. When Alexis de Tocqueville toured the United States in 1831 and 1832, he was struck by the proliferation of voluntary associations and noted that most groups welcomed participants without regard for their status or credentials. Such associations, he felt, served as crucial antidotes to the isolating tendencies of modern democratic societies and helped check the power of government. Yet according to Putnam (2000), Americans' in-person participation in voluntary associations has diminished greatly since Tocqueville's time. Even so, opportunities for youth civic engagement persist through student activities groups, community service organizations, and political parties, and many young people engage in activities through these traditional offline venues. However, participation offline requires real-time, physical presence—attending rallies, distributing leaflets, volunteering at soup kitchens, and so on. In offering new, asynchronous ways to participate and inviting everyone to have a voice, the new digital media may be contributing to a resurgence of the voluntary association model (Gardner 2007b).

Online, the role of participant is available to anyone with consistent access to the technologies that make up the new media (which are increasingly available through public libraries, schools, computer clubhouses, and so on) and to the skills (technical and social) to navigate them. Second, standards of behavior online are less explicit, and many participants resist the notion of constructing standards for fear that they would undercut freedom of expression (Pilkington 2007). Third, a distinctive feature of new media is displayed in Web 2.0, which allows online content to be modified by users. Participation is not limited to those with specific credentials and attributes (race, class, sex, age, and so on). Thus, online spaces provide opportunities to move beyond consumption and reaction to the proactive creation of content, including music, video, journalism, and identities (Floridi and Sanders 2005; Jenkins 2006a, 2006b). Proactive participation can include ethically neutral creations (such as posting a video of oneself throwing a Frisbee) as well as ethically principled ones (including political blogging, citizen journalism, and serious game design). The latter activities are forms of engagement and examples of cybercitizenship that are motivated by civic purposes, such as promoting a particular cause or viewpoint, sharing information with a broader public, and encouraging deliberation and collective problem solving.

The Promises of Online Participation

The promises of online participation are frequently touted. Benefits may come to the individual (in the form of access, acquisition of skills, sense of empowerment or efficacy, and exposure to diverse viewpoints), to the online communities themselves (through diversity of membership and information sharing), and to society (through citizen journalism, civic engagement, and democratic participation). It is not surprising that the potentials that are inherent in this virtually open public sphere have generated excitement.

First, the openness of the new digital media provides young people with opportunities to assume empowering participant roles. A young person can form and lead a film discussion group on Internet Movie Database, contribute to the creation of standards of behavior within a political discussion group on Gather .com, and become a mentor and teacher to peers and adults who are less sophisticated users of the new media. Such opportunities to assume leadership, mentoring, and educating roles can build key skills and a sense of efficacy. Furthermore, opportunities to interact with diverse participants through online dialogues, blogs, social networking sites, and massive multiplayer online games (MMOGs) can provide exposure to a wider range of ideas, opinions, and perspectives than exist in more local, offline forms of participation.

Second, regardless of attributes (such as race and sex) and formal credentials, citizens from all walks of life can contribute to the creation and distribution of knowledge and media. One positive outcome that is associated with this openness is citizen media or citizen journalism—journalism that is carried out by ordinary people who lack formal journalism training but who capture news on devices like cell phones and distribute images and text via blogs and YouTube. Decentralized news is citizen-driven and therefore local. It focuses on issues that are important to the writer (or to an intended audience) and not on the sensational headlines that news industry often relies on to sell newspapers and to attract viewers. Citizen journalism offers opportunities and skills to individuals, can enhance the quality of journalism that is produced, and thus can create a better knowledge base for deliberation about public issues. Through his video, Xander can have a voice in and contribute valuable data to a broader public dialogue about environmental degradation.

Third, opportunities for online participation mobilize young people to social and political action (Bennett 2007). According to Pettingill (2007), a new model of civic engagement, "engagement 2.0," may be emerging through the new media, spawned by the "participatory cultures" that Jenkins (2006a) suggests are starting points for a more participatory democracy. Jenkins suggests that participatory cultures are powerful because, through them, a young person can take action and make a difference.

Participation, even in spaces that are not considered political (such Facebook or *World of Warcraft*), can lead to an increased sense of efficacy, an important component of social and political engagement. This sense of efficacy contrasts sharply with the diminished sense of agency that many youth feel about traditional politics. Furthermore, as youth act through participatory cultures, they may begin to demand that traditional politics and not simply the market respond to their creations. For Jenkins (2006a, 2006b), the strong sense of community that many young people experience in these cultures may lead them to see the importance of civic ties and of their obligations to other communities of which they are members.

In sum, opportunities for youth to assume empowering social roles online can endow them with a sense of responsibility to others, to their communities, and to society. Sharing his video with a wide audience through YouTube can reaffirm Xander's perception of himself as a citizen and thus inspire further civic participation.

The Perils of Online Participation

As youth assume more proactive roles in the new digital media, a number of ethical risks can arise. First, while the new media are technically open to all, digital divides persist. Access is increasingly available in spaces such as public libraries, but some young people don't have consistent access to the new media or to support structures that guide their use or participation. To these youth, the new digital media may be viewed as intimidating instead of inviting, engaging, and empowering. A divide exists between those who have access to skills and those who don't, and youth with strong offline resources are best poised to take advantage of the participatory potentials of the new media. As skills such as multitasking, risk-taking, and mental flexibility become increasingly valuable in the workplace, nondigital youth are left behind.

Second, among young people who do participate in the new digital media, some individuals engage in hate speech, griefing, trolling (disrupting online forums or chats with offensive or off-topic posts), and other forms of misconduct online, which may be encouraged by anonymity, lack of face-to-face interaction, and the short response time of the Internet. Cyberbullying among students is on the rise, although school systems sometimes hesitate to interfere because cyberspace is outside of their purview. Far from participating as citizens with clear responsibilities, some young participants feel accountable to no one online (and one might ask, "Why should they, since they are at play?"). Although there is freedom in the absence of clear roles and responsibilities, it can result in confusion and anomie. The real or perceived absence of accountability structures means that little or no recourse exists for victims. These perils are most salient to individuals who may not perceive themselves to be members of a strong community or a "participatory culture" with a shared sense of purpose, interest, and belonging implied therein. For instance, a MySpace member might troll the site from time to time, a blogger might post his views but not perceive himself to be a citizen of the blogosphere, and a *World of Warcraft* player might have personal goals in the game that override his loyalty to his guild. In these cases, the onus is on individuals to behave in respectful and ethical ways and to respond with integrity and decisiveness when others do not. Person-centered factors (developmental stage and values) and cultural factors (peer norms) become critically important guides for behavior.

Communities themselves may dissolve if their members do not create standards of behavior and codes of conduct that are agreed on and well understood. Civility may be considered secondary if personal liberties, such as free speech, are cherished at the expense of community, as demonstrated in the controversy over the death threats that were posted on Kathy Sierra's blog. In the absence of strong ties and formal commitments to online communities, participants can join temporarily for short periods of time and roam from one community to the next. Indeed, the word *community* may not apply to spaces where membership is in continual flux and commitment is weak. In such contexts, the aforementioned opportunities for young people to work with others in building shared standards of behavior may not exist.

On the other hand, one of the most dangerous potentials related to participation in the new media is that individuals will overcommit to certain communities and fail to take advantage of the opportunity to be exposed to diverse perspectives. As users personalize their consumption of information and knowledge, balkanization and splintering can occur. Turning only to their preferred news sources, users may effectively isolate themselves from valuable alternative facts and viewpoints (Sunstein 2007). Citizen journalists may be too committed to localism at the expense of broader concerns, and citizen reporters may engage in important work but have no inherent responsibility or accountability to their communities (beyond goodwill). The citizen in *citizen media* often refers to the person contributing the media and not the citizenship or responsibilities of that citizen. Young people may limit their participation to groups that subscribe to and reinforce a myopic or prejudicial worldview. Participation in the new media can thus lead to a resurgence of hate mongering, neo-Nazi groups, or terrorist organizations as surely as it can stimulate positive deliberative discourse and the exposure of injustices large and small. Although the Internet is an impressive patchwork of diverse communities, the ways in which people participate online may preclude the dialogues across communities that constitute an authentic public sphere.

Finally, a notable peril that is related to participation in the new digital media is the frequent assumption that participatory culture is synonymous with or leads to civic engagement and democratic participation. The new digital media might hold the potential for invigorating democracy, but this doesn't mean that their potential is actually being realized. In fact, participation in these media could lead more people to withdraw from participation in real-world politics out of frustration at its inefficiencies, corruption, or remoteness from their lives. If young people see themselves as efficacious only when they're online, then they may avoid an offline political system that they already see as problematic, uninviting, and difficult to navigate. Furthermore, there may be a danger in assuming that civic engagement in virtual worlds like Second Life and MMOGs requires and engenders skills for democratic participation that are needed in the real world. Valuable lessons and skills are gained in these cyberspaces, but the transfer from online contexts to offline may not be direct (Pettingill 2007).

The Ethics of Online Participation

The new digital media's most important virtues and greatest liabilities lie in their openness. On the one hand, the new media can empower young people by inviting them to assume new empowering roles and exposing them to diverse perspectives. On the other hand, in online spaces youth can engage in bullying, avoid accountability, and circumscribe their participation to narrow interest groups. Splintering rather than greater social tolerance and responsibility is one possible outcome of participation. Whether or not they realize it, the online roles that young people are assuming—blogger, Facebook friend, filmmaker, citizen—carry responsibilities. Online participation, whether posting comments on MySpace or creating a digital film for Earth Day, involves conscious choices on the part of participants. Ethical conduct and creation online requires youth to consider carefully, as Xander did, the broader implications of personal conduct and creations. A significant onus falls on young people, but institutions and adult authority figures are also deeply implicated. Gatekeeping institutions (including local government, schools, libraries, and even families) broker initial access to technologies, while educators and other adults provide the technical skills that permit a basic level of participation and the social and ethical skills that nurture good participation.

As young people engage in the new digital media, their environment can prepare—or fail to prepare—them for the associated challenges and opportunities. On the most basic level, youth need access to technology and to the core skills that are required to use it. Ideally, access is granted in both formal and informal educational settings that are rich with traditional (older) and peer mentors. Mentors play an important role in passing on vital technical skills and in teaching young people to view themselves as participants who do not simply use media but shape it. This perspective is echoed by Jenkins et al. (2006), for whom the new media literacies entail not just traditional literacy skills (such as writing and research) but social and ethical skills as well. Youth need social skills to interact with society and to see themselves as part of it, and they need to be thoughtful and reflective about their actions. These key skills are not learned in a vacuum and certainly cannot be assumed to accompany technical skills. Here the responsibility lies with adults (parents, educators, and policymakers) to provide young people with optimal supports for good play and citizenship.

Note

1. This ordering does not suggest that the starting point—the self—is autonomous and free of social influences or effects. We begin our analysis with an issue that, on its face, may not appear to have direct ethical implications. Identity formation is primarily directed to and concerned with oneself. Yet the ways in which an individual experiments with and represents an identity online can carry ethical weight and affect others. We show that identity play can set the stage for (or overlap with) other-oriented, ethically loaded conduct that is related to issues of privacy, ownership and authorship, credibility, and participation.

Conclusion: Toward Good Play

Some are tempted to think of life in cyberspace as insignificant, as escape or meaningless diversion. It is not. Our experiences there are serious play. We belittle them at our risk. We must understand the dynamics of virtual experience both to foresee who might be in danger and to put these experiences to best use. Without a deep understanding of the many selves that we express in the virtual, we cannot use our experiences to enrich the real. If we cultivate our awareness of what stands behind our screen personae, we are more likely to succeed in using virtual experience for personal transformation. (Turkle 1995, 268)

Turkle's plea for taking virtual worlds seriously was made back in 1995 in *Life on the Screen: Identity in the Age of the Internet,* her account of multiuser, online game participants. Back then, few might have anticipated how important—indeed, routine—virtual interactions would become for many of us. Her plea resonates today across a wide spectrum of activities in which youth and adults are regularly engaged.

In this report, we provide a wide-ranging account of the ethical issues that we believe to be emerging in the new digital media. This account has been informed by interviews, emerging scholarship on new media, and theoretical insights from anthropology, cultural studies, psychology, political science, and sociology. We have concluded that ethical fault lines in the digital media revolve around five issues—identity, privacy, ownership and authorship, credibility, and participation. Our account considers evidence that "digital youth" hold distinct mental models with respect to these issues. In social networking sites, blogs, games, and other online communities that comprise the digital media, specific norms appear to be emerging around self-representation and self-expression; disclosure of personal information; creation, appropriation, and sharing of content; and conduct with others. Some of these norms—such as identity deception, either for play or for safety's sake—carry ethical stakes and suggest that distinct "ethical minds" may be emerging.

Despite the widespread participation of young people in the new digital media, little research has focused on the ethical perspectives of young people and their online pursuits. It would be unwise to presume that our largely adult-informed claims about the chief ethical fault lines in the new digital media align neatly with youth's perspectives and struggles. This report therefore is a conceptual starting point from which we—and, we hope, others—will launch empirical studies of young people themselves. We expect to revise our initial conceptualization—the themes and our understanding of their interrelationships—in light of our future research, and we expect that our studies will provide insights as to whether new frameworks of ethics are needed to address the opportunities and risks of our increasingly digital lives. Furthermore, we hope to understand whether and how traditional psychological theories of moral development may need to be revised in light of digital participation by youth at ever-younger ages.

A Model of Good Play

We define *good play* as meaningful and socially responsible participation online. The contested and evolving nature of issues such as privacy, ownership, and authorship suggest that it is premature to define what constitutes socially responsible, ethical, or good play and its opposite—irresponsible, unethical participation. Even so, certain factors are likely to contribute to a given individual's mental model or ethical stance around such issues. Our research and reflection have shown us that the ethical stances of young people are shaped by how they manage their identities and privacy, regard ownership and authorship, establish their credibility, treat others, and consider broader civic issues as they participate in online spaces. Five key sets of factors are implicated in these ethical stances (see figure 1):

• **The affordances of the new digital media** The playground of the digital media includes the technologies themselves and the structural features that invite participation and affect the likely forms that it will take. As noted and established by other scholars (e.g., Jenkins et al. 2006), Web 2.0 technologies encourage active participation. Many games and virtual worlds like Second Life invite (indeed, rely on) user contributions, such as modding of existing games. Copy and paste functionality facilitates downloading of content and information. Privacy settings on social networking sites can help users manage disclosure of personal information, yet the massive scale of the Internet can

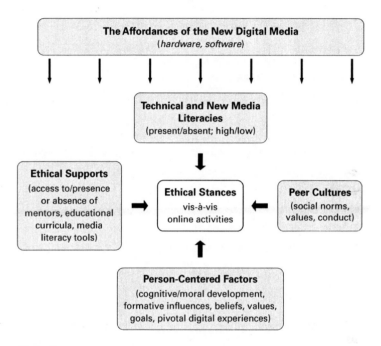

Figure 1
The Ethics of Play

create an illusory sense of privacy, safety, and anonymity that make privacy strategies seem unnecessary (Huffaker 2006). These structural features constitute the backdrop against which an individual participates and affect the likely forms that participation takes. For example, digital technologies themselves could (although most at present don't) prompt youth to consider invisible audiences; the persistence, searchability, and replicability of online information (boyd 2007b); and the negative effects of ownership and authorship transgressions. Architects of digital media hold diverse interests, including education, knowledge sharing, and profit, and whether explicit or not, these interests affect the nature of media.

• **Technical and new media literacies**   Complementing the capabilities of the new media are the skills and literacies that are required to use them effectively. The expertise of a young person can include simple knowledge of copy and paste functions, more advanced HTML programming and game design skills, and broader media literacy skills. Young people who have grown up exposed to digital technologies can navigate the Web, manipulate information and files, and artfully design their MySpace pages. The term *copy and paste literacy* is often used to describe these commonly held aptitudes. More sophisticated young people may engage in hacking and modding, some of which are illegal and some permitted and encouraged. Advanced technical abilities allow a young person to take full advantage of the new media technologies, the effects of which can be either socially positive or unethical. In short, the technologies and literacies of the new digital media—the impressive things that the technologies permit young people to do may overshadow ethical questions about what young people should do. Digital youth who possess the cognitive skills and motivation to consider the implications of their activities are well-poised to use their powers to engage in good play. Yet the acquisition of these technical, social, and ethical literacies also depends on forces (particularly as the availability of ethical supports such as mentors and new media literacy curricula) that might be outside of a young person's control.

• **Person-centered factors**   Ethical stances are also shaped by individual factors, including a young person's cognitive and moral development and the beliefs, values, and purposes that she brings to her online pursuits. For an individual to act ethically, she needs to understand possible consequences for herself, for others in her community, and for society. Such abstract thinking requires certain cognitive and moral skills, including the ability to take different perspectives, think critically about possibilities, hypothesize about the future, and make connections between actions and consequences. These skills are gained through certain kinds of experiences that often (though not always) come with age (Kegan 1994; Kohlberg 1981; Turiel 2006). With respect to ownership, for example, complex concepts such as copyright infringement may not be easily grasped by tweens, while older youth with higher-stage cognitive skills might be able to identify ethical dilemmas in authorship and privacy issues and act accordingly. As previously noted, young people are increasingly confronting these issues at relatively young ages. Despite the presence of privacy safeguards and moderators on its site, Club Penguin is not immune from problems such as cheating (Benderoff 2007).

As with all experiences, those within digital worlds can advance cognitive and moral development (Bradley 2005). Pivotal digital moments, positive or negative, are learning experiences that potentially push a young person to consider her actions in a new light and make different future choices. These moments can include empowering experiences with cocreation and participation, as well as negative experiences involving privacy lapses (oversharing and thereby harming a friend) or identity play that deceives and harms another person. A young person can assume different roles within such scenarios (such as intentional or unintentional perpetrator, victim, or bystander) and still gain insights that further his or her moral skills. Notwithstanding these valuable learning experiences, a digitally savvy child (or tween) cannot be expected to grasp all of the possible ethical consequences of her choices in digital publics.

Equally important to cognitive and moral capacities are the more stable beliefs and values held by a young person, which may stem from the formative influences of family, religion, and other sources. Additionally, a strong sense of purpose, as exemplified by our environmentalist Xander, can engender (but by no means guarantee) ethical participation (Bazerman 2006).

▪ **Peer cultures**   Both online and offline peer norms and values constitute powerful influences for youth. Our analysis refers to an online infringing youth culture, meaning that young digital media users may have a sense of entitlement about information and property that normalizes illegal downloading and thus may infringe on the rights of musicians and other creators. Youth may feel justified in such illicit activities if their own ownership claims are ignored (as in the case of game modders) or may define their actions as robbing the rich as Robin Hood did.

We also discuss evidence for the existence of youth cultures of disclosure that encourage and reward the sharing of personal information on social networking sites and blogs, often aided by practices such as selective disclosure, deception, or code switching. Peer norms on social networking sites are powerful. Young people study other Facebook users' profiles, note the kinds of details disclosed, and often model their own profiles after the appealing models. Social networking sites are becoming important spaces for the transmission of cultural tastes through favorites lists (Jason Kaufman, personal communication, March 22, 2007). The desire to be accepted is a powerful incentive for mimicking other youth's profiles and revealing personal information, often without considering potentially negative consequences. At the same time, socially responsible cultures also exist online, and groups with explicit good play agendas have emerged. Through Teen Angels, youth educate peers about cyberbullying and encourage responsibility on privacy and predator issues. Youth Radio and Global Kids engage in civic pursuits online, build educational games, and produce citizen journalism and are thus positive role models of youth participation.

Second, offline peer cultural influences may be as important as the technologies, literacies, and youth cultures in the digital media. As established by previous research, social pressures on young people to succeed, the absence of mentors, and the presence of a peer culture that condones cheating can lead to unethical conduct (Callahan 2004; Fischman et al. 2004; Josephson Institute of Ethics 2006). Youth cultures that are dedicated to good play exist offline, and through sports, community service, and student political groups, young people can and do model ethical conduct. However, the cheating culture may be equally powerful, especially given the noted mounting evidence about the absence of adult ethical supports.

• **Ethical supports** Adult supports—parent role models, teacher mentors, and school curricula—can play decisive roles in young people's online choices. Positive adult role models can provide resources to help youth buck the norms of the offline cheating culture and make considered choices online with respect to identity, privacy, ownership and authorship, credibility, and participation. If a young person's parents engage in software piracy, they unwittingly reinforce the norms of the infringing culture. If there are few digital mentors (individuals with greater technical and ethical knowledge and experience), then a young person may have no supports for reflecting on the larger implications, for himself and others, of sharing details of his college drinking adventures on his MySpace page. If schools limit on-campus access to certain Web sites and yet fail to provide students with the literacies that they need to navigate the frontier of the Web out of school, then they are doing little to prevent unethical conduct. New media literacies curricula can go a long way toward encouraging good play but require adoption and promotion by adult educators.

Ethical supports can also be and increasingly are provided through digital media themselves. Educational games like *Quest Atlantis* and curricula such as the New Media Literacies Learning Library (both of which are available online for anyone to access) prompt participants to consider ethical issues, but their ethical lessons may be better grasped if the online experience is supplemented by offline adult or peer reflection and discussion. At the same time, commercial entities have an increasing presence on the sites that youth most frequent (such as Facebook and My-Space), and industry may be supporting—or undermining—critical thinking about privacy, identity, and other issues discussed above. On the whole, it seems urgent to consider which stakeholders—education, industry, or government—are best poised to define the public interest, to lead conversations about digital ethics, and to scaffold young people around these issues.

Ideally, our good play model provides a balance of technologies, opportunities, and support that set the stage for young people to become productive, innovative, and ethical participants in the new digital media. At present, however, the burden of good play largely falls on individual youth. The frontiers of the new digital media permit and empower youth to engage in largely free play and participate in the public sphere in new ways and to an unprecedented extent. The structures of the technologies themselves set few limitations, and in this there are both tremendous promises and significant perils for young people. At the same time, evidence suggests that detrimental peer cultures exist (and may be more powerful than socially responsible cultures) and that ethical supports (mentors, role models, and educational curricula) may be rare (Fischman et al. 2004; Josephson Institute of Ethics 2006). There is tremendous pressure on young people to develop the cognitive and moral skills and integrity of beliefs, values, and purposes that engender good play.

Research on Good Play: The Need for Deeper Empirical Study

The proposed model of good play in the new digital media sets the stage for an empirical study that invites young people to share their stories and perspectives about their digital lives. To explore the extent to which our treatment of ethical fault lines aligns with youth attitudes, conduct, and experiences with others, our research will explore the following questions:

- What mental models do young people hold about online ethical issues?

- How do they think about the ethical connotations of their digital media play?
- What variations exist among youth in their ethical approaches to the new media?
- What are the leading areas of confusion and inconsistency?

We need to understand how and under what circumstances privacy and credibility are experienced by youth as ethical issues and in what situations young people believe that appropriating online content is ethical versus unethical. We hope to learn the extent to which ethical supports exist for average youth as they participate in the new digital media. Overall, we seek to understand how person-centered factors interplay with the technologies of the digital media, technical and new media literacies, peer cultures, and ethical supports in affecting how a young person conceives of (and engages in) good play. We have some suspicions but plan to proceed in eliciting the perspectives of young people themselves before making definitive conclusions about the ethical fault lines at play. We will conduct qualitative interviews that explore the everyday activities of young people and, from their point of view, the salient ethical issues that come up, the ways that they manage them, and the supports that guide their choices.

Interventions and Supports for Good Play: The Need for Research-Based Interventions

As young people immerse themselves in digital environments, they need to be equipped with the capacities to act responsibly there. Ultimately, our research efforts are motivated by a desire to create ethical supports for young people to reflect about what constitutes good play—meaningful and socially responsible pursuits—both online and off. Countless examples of ethical misconduct and confusion online suggest a pressing need. For the promises of the new digital media to be positively realized, supports for the development of ethical skills—or, better yet, "ethical minds" (Gardner 2007a)—must emerge. Although it is clear that a complex set of factors is producing the ethical stances that young people hold in relation to their online activities, encouraging them to reflect on these issues can be an important intervention. Youth who consider their roles in various online contexts, understand the responsibilities that are implied by them, and imagine the larger implications of various judgments, are well-poised to engage in good play.

With Henry Jenkins and his Project New Media Literacies team, we are developing prototypes of curricular exercises that are designed to meet these objectives. The curriculum places a premium on role-playing activities that bring to light and ask participants to confront ethical issues that are raised in the new media landscape. Such role-playing exercises will be buttressed by case examples of "real" ethical problems that are discussed by our youth interview participants and by professional media makers in the video interviews that make up the New Media Exemplar Library produced by Jenkins's team. The final product will be comprised of five or more modules, each organized around a central ethical issue—the five issues considered above and perhaps other yet-to-be discovered issues that surface in our research.

We seek to understand and encourage good play not to create more obedient, respectful youth but to develop ethical reflection and conduct as a key foundation for youth empowerment. The new digital media create tremendous opportunities for young people—to nurture important skills, to connect with others around the world, to engage in meaningful play, to nurture skills for future careers, to engage in civic pursuits, and to contribute to a greater good. Our hope is that our work helps to cultivate these promises while minimizing the risks that lie in the frontiers of digital media.

Appendix A: Youth Engagement with the New Digital Media

Young people today are frequently engaged in the following activities—and thus assume a number of different roles—through the new media:

- **Self-expression and identity experimentation** These activities include creating avatars through role-playing games and virtual worlds; creating and sharing content (text, music, and video) individually and collaboratively through blogs (LiveJournal, Xanga), vlogs (YouTube), and music sharing sites (MySpace). Studies suggest that 57 percent of online teens create content, including blogs (Lenhart and Madden 2005), and even younger children are increasingly playing active, creator roles online (Green and Hannon 2007).
- **Social networking** These activities include chatting with friends, reaching out to people with shared interests, and establishing support groups (Facebook, MySpace). According to a recent Pew study, 55 percent of online teens use social networks and have created online profiles, 91 percent of teens chat with offline friends through these sites, and half pursue new online friendships (Lenhart and Madden 2007).

- **Gaming** These activities include single-player and multi-player, role-playing games (such as *World of Warcraft*). Gaming is a popular youth activity. The average thirteen- to eighteen-year-old plays fourteen hours of video games per week (Martin and Oppenheim 2007), and over half of the 117 million "active gamers" in the United States play games online (Graft 2006).

- **Consumption and entertainment** These activities include downloading music (iTunes), watching videos (YouTube), and shopping (Amazon). Pew's 2005 study of online content found that half of online teens download music (Lenhart and Madden 2005).

- **Educating** These activities include teaching and mentoring others (for example, with technical skills and online game strategies). Through programs such Youth Radio, Computer Clubhouses, Scratch, online gaming communities, and other informal learning environments, young people are increasingly learning with and from their peers technical skills and game strategies.

- **Knowledge-building** These activities include research, school work, news, and other information gathering (including Wikipedia, Google, and NYTimes.com). According to Pew's recent report on Wikipedia, young adults are more likely than older adults to turn to Wikipedia. Forty-four percent of those ages eighteen to twenty-nine turn to Wikipedia for information, compared to only 29 percent of users age fifty and older (Rainie and Tancer 2007).

- **Dialogue and civic engagement** These activities include engaging in public discourse, promoting social change, and political, social, and cultural criticism. Through programs such as Youth

Radio and the Global Kids Online Leadership Program and sites such as Gather.com, young people are educating their peers about key social issues, and mentoring civic engagement and activism online.

Appendix B: Informant Interview Protocol

The following general template of questions was used as a starting point in preparing for interviews with informants. In each interview, questions were tailored to the background and expertise of the specific informant.

The GoodPlay Project: Ethical Perspectives on Youth and Digital Media

Informant Interview Protocol (General Template)

I. Broad entry questions
1. Can you tell us how you became interested in researching / teaching youth / or participating in the new digital media?
   a. What findings from this research have been most surprising or intriguing to you?
   b. What is the focus of your current and future research?
II. Digital media: buckets, goals, and roles of participants
2. How would you define the domain of digital media?
   a. How would you parse the domain? In other words, what are

the most important "buckets" (or major types of activities) that make up the domain?

3. Which buckets of the digital media are most important to explore in a study of young people?

   a. If applicable: What kinds of digital activities are the kids you studied most frequently engaged? What specific sites do they frequent (MySpace, Facebook, YouTube, Second Life, *World of Warcraft*, and so on)?

4. What are the various goals of participants in these activities?

   a. Is there consensus around the goals of participation in a given space (MySpace, games, blogging, and so on)?

   b. Have you witnessed instances when the goals or values of participants are in conflict? (Example: A jokester "crashes" a massive multiplayer online game, pretending at first to be a serious player, winning the trust of coplayers, and then undermining the game at an opportune moment). If so, how was the conflict resolved?

   c. Can you think of a case (or space) in which the conflicting goals or values of participants were successfully managed? How was this accomplished?

5. What kinds of roles are these young people playing in these spaces?

   a. Are these roles explicitly defined?

   b. What kinds of responsibilities accompany these roles?

   c. Are these responsibilities explicitly acknowledged or implicit?

III. Ethical issues

6. In your experience studying / teaching youth about / participating in the digital media, have you come across

situations in which youth (or adults) struggle over right versus wrong courses of action?

    a. In other words, what types of ethical dilemmas have you come across?

    b. Are these dilemmas unique to the digital space?

    c. Are there distinct ethical situations or dilemmas that arise among young participants? Describe.

7. Do distinct ethical issues emerge in the different buckets that make up the domain?

    a. For example, what kinds of ethical issues and dilemmas are common in the blogging space? In multiuser games? In online communities? In chat rooms?

    b. Are any of these issues unique to a particular bucket or to the online (versus offline) world?

8. When there is unethical behavior (or behavior that is seen as unethical), what sanctions are imposed? By whom?

9. How aware are young people of the ethical implications of their online conduct?

    a. In your research, did you find evidence of awareness of the ethical implications of one's conduct online? Do specific examples or incidents that reflect such awareness come to mind?

    b. Are there ethical issues relating to the Internet that you believe young people in particular are unaware of or deliberately ignore? Do specific examples or incidents come to mind?

    c. Are there ethical issues that you think young people should be made aware of? If yes, do you have any ideas about how this could be best accomplished?

10. Are the ethical concerns (and awareness) of young people similar to or qualitatively different from those of older generations? If different, how?

11. Broadly speaking, what major ethical concerns do you have about the digital media?

IV. Mentors

12. Based on your knowledge of this space, do you have a sense of whom kids turn to for advice in their activities online? Do they have mentors?

   a. If yes, who are they (peers versus traditional mentors or individuals with greater in age, experience, or wisdoms)?

   b. Some would argue that peer mentoring is more common for youth participants in new media. How is peer mentoring different from (and similar to) traditional mentoring in this space? Where, when, and how does digital mentoring happen?

   c. What are the implications of peer mentoring for awareness of ethical issues and for encouraging ethical conduct? In other words, do you think that peer mentors are capable of instilling ethics in their mentees in the same way that traditional mentors do in other domains?

13. Is there evidence that kids have "antimentors" or well-developed conceptions of the kind of conduct online that is inappropriate, disrespectful, and so on? If yes, elaborate.

V. General opportunities and challenges of the new digital media

14. What are the greatest opportunities offered by the Internet? For young people?

   a. Do you think that the Internet opens up unique opportunities for civic engagement? If yes, could you describe how? If no, why not?

      b. Do you think the greatest opportunities of the Internet can or will be realized? If so, when and how? If not, what obstacles might prevent their realization?

   15. What are the greatest challenges posed by the Internet? For young people?

      a. What are your thoughts on the digital divide between white middle-class kids and less privileged kids? Do you perceive this gap to be closing?

      b. Do you think these challenges can or will be surmounted? If so, when and how? If not, why not?

VI. Ethical issues in digital research

   16. What major challenges can you foresee for us in conducting this research?

   17. Can you speak generally about any major ethical considerations in doing research on the digital media that we should bear in mind as we go forward?

   18. In addition to conducting interviews such as this, we hope to observe young people as they engage in various online interactions. For instance, other researchers (developmental psychologists) who study the social interactions of adolescents on the Internet have entered teen chat rooms as passive observers.

      a. What are your reactions to this?

      b. Do you see any ethical issues involved in this type of research?

      c. What might be some alternative ways of learning about how kids are interaction online?

VII. Conclusion / information gathering

   19. Is there anything relevant to digital media, ethics, and young persons that you would like to add that I didn't ask you about?

20. Can you recommend other individuals with whom we should speak (including other educators working with kids and technology, experts, researchers, as well as very experienced participants – both youth and adult)?

References

Ad Council. 2007. "PSA Campaign: Online Sexual Exploitation." Retrieved November 20, 2007, from http://www.adcouncil.org.

Anderson, C. A., N. L. Carnagey, M. Flanagan, A. J. Benjamin, J. Eubanks, and J. C. Valentine. 2004. "Violent Video Games: Specific Effects of Violent Content on Aggressive Thoughts and Behavior." *Advances in Experimental Social Psychology* 36: 199–249.

Bateson, G. 1972. *Steps to an Ecology of Mind.* San Francisco: Chandler.

Bazerman, M. H. 1990. *Judgment in Managerial Decision Making.* New York: Wilcy.

BBC News. 2007. "Students Assessed with Wikipedia." BBC News, March 6. Retrieved December 11, 2007, from http://news.bbc.co.uk/go/pr/fr/-/2/hi/uk_news/education/6422877.stm.

Benderoff, E. 2007. "Cheating a Real Problem in Club Penguin's Virtual World." *Chicago Tribune.* Retrieved February 1, 2008, from http://www.chicagotribune.com/business/chi-0703080167mar08,0,4256114story?coll=chi-bizfont-hed.

Bennett, W. L. 2007. "Changing Citizenship in the Digital Age." In *Civic Life Online*, ed. L. W. Bennett, 1–24. Cambridge, MA: MIT Press.

Blogger's Code of Conduct. 2007. Retrieved April 24, 2007, from http://radar.oreilly.com/archives/2007/04/draft-bloggers-1.html.

Bok, S. 1979. *Lying: Moral Choice in Public and Private Life.* New York: Vintage Books.

Bosman, J. 2006. "The Kid with All the News about TV News." *New York Times,* November 20. Retrieved December 28, 2006, from http://travel.nytimes.com/2006/11/20/business/media/20newser.html.

boyd, d. 2007a. "The Cost of Lying." *Apophenia,* January 8. Retrieved January 24, 2007, from http://www.zephoria.org/thoughts/archives/2007/01/08/thecost_of_lyi.html.

boyd, d. 2007b. "Why Youth (Heart) Social Networking Sites: The Role of Networked Publics in Teenage Social Life." In *Youth, Identity and Digital Media,* ed. D. Buckingham, 119–142. Cambridge, MA: MIT Press.

boyd, d., and J. Heer. 2006. "Profiles as Conversation: Networked Identity Performance on Friendster." In *Proceedings of the Hawaii International Conference on System Sciences (HICSS-39).* Kauai, HI: IEEE Computer Society. Retrieved April 16, 2009, from http://www.danah.org/papers/HICSS2006.pdf.

Bradley, K. 2005. "Internet Lives: Social Context and Moral Domain in Adolescent Development." *New Directions for Youth Development* 108: 57–76.

Buckingham, D. 2000. *After the Death of Childhood: Growing Up in the Age of Electronic Media.* Malden, MA: Blackwell.

Buckingham, D. 2003. *Media Education: Literacy, Learning and Contemporary Culture.* Cambridge, MA: Polity Press.

Buckingham, D. 2007. "Introducing Identity." In *Youth, Identity, and Digital Media,* ed. D. Buckingham, 1–22. Cambridge, MA: MIT Press.

Callahan, D. 2004. *The Cheating Culture: Why More Americans Are Doing Wrong to Get Ahead.* Orlando: Harcourt.

Chavez, V., and E. Soep. 2005. "Youth Radio and the Pedagogy of Collegiality." *Harvard Educational Review* 75, no. 4: 409–434.

Cooley, C. H. 1902. *Human Nature and the Social Order*. New York: Scribner's.

Davidson, C. N. 2007. "We Can't Ignore the Influence of Digital Technologies." *Chronicle of Higher Education* 53, no. 2: B20–B20.

de la Merced, M. J. (2006, October 21). "A Student's Video Résumé Gets Attention (Some of It Unwanted)." *New York Times,* October 21. Retrieved June 10, 2007, from http://www.nytimes.com/20061021/business/21bank. html?ex=1181620800&en=2e3aef2615719766&ei=5070.

Deleting Online Predators Act. 2006. Retrieved April 26, 2007, from http://www.govtrack.us/congress/bill.xpd?bill=h109-5319.

Digital Millennium Copyright Act of 1998. 1998. http://frwebgate. access.gpo.gov/cgi-bin/getdoc.cgi?dbname=105_cong_public_ laws&docid=f:publ304.105.pdf.

Digital Natives. 2007. "What Is the Digital Natives Project?" Retrieved December 10, 2007, from http://www.digitalnative.org.

Donath, J. 1999. "Identity and Deception in the Virtual Community." In *Communities in Cyberspace*, ed. M. A. Smith and P. Kollock, 29–59. New York: Routledge.

Edelman. 2007. "Edelman Trust Barometer." Retrieved June 30, 2009, from http://www.edelman.com/trust/2007/trust_final_1_31.pdf.

Erikson, E. H. 1968. *Identity, Youth, and Crisis*. New York: Norton.

Erikson, E. H. 1980. *Identity and the Life Cycle*. New York: Norton.

Fischman, W., B. Solomon, D. Greenspan, and H. Gardner. 2004. *Making Good: How Young People Make Moral Decisions at Work*. Cambridge, MA: Harvard University Press.

Floridi, L., and J. W. Sanders. 2005. "Internet Ethics: The Constructionist Values of Homo Poieticus." In *The Impact of the Internet on Our Moral Lives*, ed. R. Cavalier, 195–214. Albany: SUNY Press.

Gardner, H. 2007a. *Five Minds for the Future*. Cambridge, MA: Harvard Business School Press.

Gardner, H. 2007b. "The Unlimited Frontiers." MacArthur Spotlight blog, April 10. Retrieved June 10, 2007, from http://spotlight.macfound .org/main/entry/unlimited_frontiers.

Gardner, H., M. Csikszentmihalyi, and W. Damon. 2001. *Good Work: When Excellence and Ethics Meet*. New York: Basic Books.

Gee, J. P. 2003. *What Video Games Have to Teach Us about Learning and Literacy*. New York: Palgrave Macmillan.

Gee, J. P. 2004. *Situated Language and Learning: A Critique of Traditional Schooling*. New York: Routledge.

Geertz, Clifford. 1972. "Deep Play: Notes on the Balinese Cockfight." In *Interpretation of Cultures*, 412–453. New York: Basic Books.

Gentile, D. A., P. J. Lynch, J. R. Linder, and D. A. Walsh. 2004. "The Effects of Violent Video Game Habits on Adolescent Hostility, Aggressive Behaviors, and School Performance." *Journal of Adolescence* 27: 5–22. Retrieved May 15, 2007, from Ebsco.

Glass, J. 1993. *Shattered Selves: Multiple Personality in a Postmodern World*. Ithaca, NY: Cornell University Press.

Goffman, E. 1959. *The Presentation of Self in Everyday Life*. New York: Doubleday.

Graft, K. 2006. "Nielson Study Reveals Gamer Habits." *Next Generation*, October 5. Retrieved June 7, 2007, from http://www.next-gen.biz/index .php?option=com_content&task=view&id=3947&Itemid=2.

Green, H., and C. Hannon. 2007. *Their Space: Education for a Digital Generation*. London: Demos.

Green, J. and H. Jenkins. 2009. "The Moral Economy of Web 2.0: Audience Research and Convergence Culture." In *Media Industries: History,*

Theory and Methods, ed. J. Holt and A. Perren, 213–226. Malden, MA: Blackwell.

Grossman, L. 2006. "Time's Person of the Year: You." *Time*, December 13. Retrieved April 26, 2007, from http://www.time.com/time/magazine/article/0.9171,1569514,00.html.

Hall, G. S. 1904. *Adolescence: Its Psychology and Its Relations to Physiology, Anthropology, Sociology, Sex, Crime, Religion and Education*. New York: Appleton.

Heffernan, V., and T. Zeller. 2006. "'Lonely Girl' (and Friends) Just Wanted Movie Deal." *New York Times*, September 12. Retrieved June 10, 2007, from http://www.nytimes.com/2006/09/12/technology/12cnd-lonely.hteml?ex=1315713600&en=abf28fc073b3c6e9&ei=5088&partner= rssnyt&emc=rss.

Huffaker, D. 2006. "Teen Blogs Exposed: The Private Lives of Teens Made Public." Paper presented at the meeting of the American Association for the Advancement of Science, St. Louis, Missouri, February 16–19.

Huizinga, J. 1955. *Homo Ludens: A Study of the Play-Element in Culture*. Boston: Beacon Press.

Ito, J. 2004. *Weblogs and Emergent Democracy*. Twenty-first Chaos Communication Congress. Retrieved April 16, 2009, from http://90.146.8.18/en/archiv_files/20041/FE_2004_joichiito_en.pdf.

Jenkins, H. 2006a. *Convergence Culture: Where Old and New Media Collide*. New York: New York University Press.

Jenkins, H. 2006b. *Fans, Bloggers, and Gamers: Media Consumers in a Digital Age*. New York: New York University Press.

Jenkins, H., K. Clinton, R. Purushotma, A. J. Robison, and M. Weigel. 2006. *Confronting the Challenges of Participatory Culture: Media Education for the Twenty-first Century*. MacArthur Foundation. Retrieved April 26, 2007, from http://www.digitallearning.macfound.org/site/c.enJLKQNlFi G/b,2108773,0240-4714.

Johnson, Steven. 2005. *Everything Bad Is Good for You*. New York: Riverhead Books.

Josephson Institute of Ethics. 2006. *2006 Report Card on American Youth*. Press release. Retrieved May 29, 2007, from http://www .josephsoninstitute.org/pdf/ReportCard_press-release_2006-1015.pdf.

Keen, A. 2007. "Can We Save the Internet? The WWW Is a Scary Introduction to Primeval Man." May 29. http://www.jewcy.com/ dialogue/2007-05-29/can_the_internet_be_saved.

Kegan, R. 1994. *In Over Our Heads: The Mental Demands of Modern Life*. Cambridge, MA: Harvard University Press.

Kohlberg, L. 1981. *Essays on Moral Development*. San Francisco: Harper and Row.

Lembo, R. 2000. *Thinking through Television*. New York: Cambridge University Press.

Lenhart, A., and M. Madden. 2005. *Teen Content Creators and Consumers*. Pew Internet and American Life Project. November 2. Retrieved June 10, 2007, from http://www.pewinternet.org/pdfs/PIP_Teens_Content_ Creation.pdf.

Lenhart, A., and M. Madden. 2007. *Social Networking Websites and Teens*. Pew Internet and American Life Project. January 7. Retrieved June 10, 2007, from http://www.pewinternet.org/pdfs/PIP_SNS_Data_Memo_Jan_ 2007.pdf.

Lessig, L. 2004. *Free Culture: How Big Media Uses Technology and the Law to Lock Down Culture and Control Creativity*. New York: Penguin Press.

Lévy, Pierre. 1999. *Collective Intelligence: Mankind's Emerging World in Cyberspace*. Trans. Robert Bonomo. New York: Basic Books.

Livingstone, S. M. 2002. *Young People and New Media: Childhood and the Changing Media Environment*. Thousand Oaks, CA: Sage.

"Lonelygirl15." 2007. Wikipedia. Retrieved June 10, 2007, from http://en.wikipedia.org/w/index.php?title=Lonelygirl15&oldid=137300907.

Lynn, R. 2007. "Virtual Rape Is Traumatic, but Is It a Crime?" *Wired*, May 4. Retrieved June 20, 2007, from http://www.wired.com/culture/lifestyle/commentary/sexdrive/2007/05/sexdrive_0504.

Madden, M., S. Fox, A. Smith, and J. Vitak. 2007. *Digital Footprints: Online Identity Management and Search in the Age of Transparency*. Pew Internet and American Life Project. Retrieved December 17, 2007, from http://www.pewintern.org/jpdfds/PIP_Digital_Footprints.pdf.

Markus, H., and P. Nurius. 1986. "Possible Selves." *American Psychologist* 41, no. 9: 954–969.

Martin, S., and K. Oppenheim. 2007. "Video Gaming: General and Pathological Use." *Trends and Tudes* 6, no. 3: 1–6.

Mead, G. H. 1934. *Mind, Self, and Society, from the Standpoint of a Social Behaviorist*. Chicago: University of Chicago.

Moore, J. F. 2003. *The Second Superpower Rears Its Beautiful Head*. Berkman Center for Internet and Society, Harvard University. Retrieved June 10, 2007, from http://cypber.law.harvard.edu/people/jmoore/secondsuperpower.pdf.

Moser, J. 2007. "So, How Am I Doing?" *New York Times*, May 1. Retrieved May 3, 2007, from http://thegraduates.blogs.nytimes.com/2007/05/01/so-hoow-am-i-doing/?ex=1178769600&en=9d6fde2b6fbc5cbb&ei=5070&emc=eta1.

Moshman, D. 2005. *Adolescent Psychological Development: Rationality, Morality, and Identity* (2nd ed.). Mahwah, NJ: Erlbaum.

MySpace. 2007. "MySpace Terms of Service." Retrieved June 18, 2007, from http://www.myspace.com/index.cfm?fuseaction=misc.terms.

O'Reilly, T. 2005. "What Is Web 2.0? Design Patterns and Business Models for the Next Generation of Software." September 30. http://www.oreillynet.com/pub/a/oreilly/tim/news/2005/09/30/what-is-web-20.html.

Pettingill, L. 2007. *Engagement 2.0? How the New Digital Media Can Invigorate Civic Engagement*. GoodWork Project Paper Series, No. 50. Retrieved November 17, 2007, from http://www.pz.harvard.edu/ebookstore/detail.cfm?pub_id=391.

Pilkington, E. 2007. "Howls of Protest as Web Gurus Attempt to Banish Bad Behaviour from Blogosphere." *The Guardian*, April 10. Retrieved June 10, 2007, from http://technology.guardian.co.uk/news/story/0,,2053278,00.html.

Postigo, H. 2003. "From Pong to Planet Quake: Post-industrial Transitions from Leisure to Work." *Information, Communication, and Society* 6, no. 4: 593–607.

Prensky, M. 2001. "Digital Natives, Digital Immigrants." *On the Horizon* 9, no. 5: 1–5.

Putnam, R. 2000. *Bowling Alone: The Collapse and Revival of American Community*. New York: Simon and Schuster.

Putter, J. 2006. "Copyright Infringement V. Academic Freedom on the Internet: Dealing with Infringing Use of Peer to Peer Technology on Campus Networks." *Journal of Law and Policy* 14: 419–470.

Radway, J. 1985. *Reading the Romance: Women, Patriarchy, and Popular Culture*. Chapel Hill: University of North Carolina.

Rainie, L., and B. Tancer. 2007. *Wikipedia Users*. Pew Internet and American Life Project. April 24. Retrieved June 10, 2007, from http://www.pewinternet.org/pdfsd/PIPP_Wikipedia07.pdf.

Rundle, M., and C. Conley. 2007. *Ethical Implications of Emerging Technologies: A Survey*. Paris: UNESCO.

Saslow, E. 2007. "Teen Tests Internet's Lewd Track Record." *Washington Post*, May 29, A01.

Schwartz, S. J. 2001. "The Evolution of Eriksonian and Neo-Eriksonian Identity Theory and Research: A Review and Integration." *Identity* 1, no. 1: 7–58.

Shaffer, D. 2006. *How Computer Games Help Children Learn*. New York: Palgrave MacMillan.

Silverstone, R. 2007. *Media and Morality: On the Rise of the Mediapolis*. Cambridge, MA: Polity Press.

Stern, S. 2007. "Producing Sites, Exploring Identities: Youth Online Authorship." In *Youth, Identity and Digital Media*, ed. D. Buckingham, 95–117. Cambridge, MA: MIT Press.

Sunstein, C. R. 2007. *Republic.com 2.0*. Princeton, NJ: Princeton University Press.

Swains, H. 2007. "Dying for Attention: Why People Are Killing Themselves Online." *Columbia News Service*. February 27. Retrieved April 16, 2009, from http://jscms.jrn.columbia.edu/cns/2007-02-27/swains-fakingdeath.

Thompson, E. P. 1971. "The Moral Economy of the English Crowd in the Eighteenth Century." *Past and Present* 50: 76–136.

Turiel, E. 2006. "The Development of Morality." In *Handbook of Child Psychology*, Vol. 3, *Social, Emotional, and Personality Development*, ed. W. Damon and R. M. Lerner, 789–857. Hoboken, NJ: Wiley.

Turkle, S. 1995. *Life on the Screen: Identity in the Age of the Internet*. New York: Simon and Schuster.

Turkle, S. 1999. "Cyberspace and Identity." *Contemporary Sociology* 28, no. 6: 643–648.

Turkle, S. 2004. "Whither Psychoanalysis in Computer Culture?" *Psychoanalytic Psychology: Journal of the Division of Psychoanalysis* 21 (Winter): 16–30.

Turkle, S. 2008. "Always-on/Always-on-You: The Tethered Self." In *Handbook of Mobile Communications and Social Change*, ed. J. Katz, 121–138. Cambridge, MA: MIT Press.

Valentine, G., and S. Holloway. 2002. "Cyberkids? Exploring Children's Identities and Social Networks in On-line and Off-line Worlds." *Annals of the Association of American Geographers* 92, no. 2: 302–319.

Vatican. 2002. "Ethics in Internet." Retrieved April 16, 2009, from http://www.vatican.va/roman_curia/pontifical_councils/pccs/documents/rc_pc_pccs_doc_20020228_ethics-internet_en.html.

Warren, S. D., and L. D. Brandeis. 1890. "The Right to Privacy." *Harvard Law Review* 4: 193–223.

Weber, S. 2006. "Sandra Weber Thinks That 'Public' Is Young People's New 'Private.'" MacArthur Spotlight blog, December 12. Retrieved December 14, 2006, from http://spotlight.macfound.org/main/entry/Sandra_weber_public_young_people_private.

Williams, R. 1974. *Television: Technology and Cultural Form*. London: Routledge.

Woo, J. 2006. "The Right Not to Be Identified: Privacy and Anonymity in the Interactive Media Environment." *New Media and Society* 8, no. 6: 949–967.

YouTube. 2007. "Terms of Service." Retrieved June 20, 2007, from http://www.youtube.com/t/terms.

Zaslow, J. 2007. "The Most-Praised Generation Goes to Work." *Wall Street Journal* (Eastern ed.), April 20, W1.

Young People, Ethics, and the New Digital Media

This report was made possible by grants from the John D. and Catherine T. MacArthur Foundation in connection with its grant making initiative on Digital Media and Learning. For more information on the initiative visit www.macfound.org.

The John D. and Catherine T. MacArthur Foundation Reports on Digital Media and Learning

The Future of Learning Institutions in a Digital Age by Cathy N. Davidson and David Theo Goldberg with the assistance of Zoë Marie Jones

New Digital Media and Learning as an Emerging Area and "Worked Examples" as One Way Forward by James Paul Gee

Living and Learning with New Media: Summary of Findings from the Digital Youth Project by Mizuko Ito, Heather Horst, Matteo Bittanti, danah boyd, Becky Herr-Stephenson, Patricia G. Lange, C. J. Pascoe, and Laura Robinson with Sonja Baumer, Rachel Cody, Dilan Mahendran, Katynka Z. Martínez, Dan Perkel, Christo Sims, and Lisa Tripp

Young People, Ethics, and the New Digital Media: A Synthesis from the Good-Play Project by Carrie James with Katie Davis, Andrea Flores, John M. Francis, Lindsay Pettingill, Margaret Rundle, and Howard Gardner

Confronting the Challenges of Participatory Culture: Media Education for the 21st Century by Henry Jenkins (P.I.) with Ravi Purushotma, Margaret Weigel, Katie Clinton, and Alice J. Robison

The Civic Potential of Video Games by Joseph Kahne, Ellen Middaugh, and Chris Evans